Introduction to Virology

Introduction to Virology

Editor: Paul Mahoney

Callisto Reference

www.callistoreference.com

Callisto Reference,
118-35 Queens Blvd., Suite 400,
Forest Hills, NY 11375, USA

Visit us on the World Wide Web at:
www.callistoreference.com

ISBN: 978-1-64116-006-3 (Hardback)

Printed in China.

Cataloging-in-Publication Data

Introduction to virology / edited by Paul Mahoney.
p. cm.
Includes bibliographical references and index.
ISBN 978-1-64116-006-3
1. Virology. 2. Microbiology. I. Mahoney, Paul.
QR360 .I58 2018
616.910 1--dc23

Table of Contents

Preface

The study of viruses is known as virology. It focuses on the structure, evolution and behavior of viruses. Studying them is vital, as they cause various infectious diseases like dengue, yellow fever, smallpox, etc. The classification of viruses is done on the basis of the host that they infect, like fungal viruses, bacteriophages, animal viruses, etc. This book attempts to assist those with a goal of delving into the field of virology. Coherent flow of topics, student-friendly language and extensive use of examples make this textbook an invaluable source of knowledge.

A detailed account of the significant topics covered in this book is provided below:

Chapter 1- Virus is an agent that is harmful in nature and can replicate and multiply when inside an organism. Virology is the study of these viruses, their structure, evolution and their influence on the host body. According to shapes, viruses can be classified into icosahedral and helical symmetries. The chapter on virus offers an insightful focus, keeping in mind the complex subject matter.

Chapter 2- Virus comprises of genomes, proteins and lipids. Genomes contain either DNA or RNA, which helps in synthesis of viral proteins and the replication process. Since viruses require a host body to replicate, they are isolated using methods of cell culture, embryonated eggs and small laboratory animals. Out of these methods, cell culture is mostly preferred for the process. This section has been carefully written to provide an easy understanding of the varied facets of general virology.

Chapter 3- Viruses were initially classified on the basis of the diseases they caused or on their geographical location. However, this changed after the introduction of the Baltimore classification. It categorised the viruses on the basis of their genome and replication methods. Classification of viruses and positive strand RNA virus is best understood in confluence with the major topics listed in the following chapter.

Chapter 4- Negative sense RNA viruses are a part of the Mononegavirales order. Owing to its negative polarisation, these viruses need to be converted into positive RNA and therefore is not infectious by themselves. Its genome has features like linear non-segmented negative sense RNA genome, the leader RNA is neither capped nor polyadenylated, and its helical nucleocapsid contains the RNA dependent RNA polymerase among other features. The major categories of negative sense RNA viruses are dealt with great details in the section.

Chapter 5- A major known RNA virus is double-stranded RNA viruses, which can acquire and survive in a range of hosts including humans, animals, fungi, plants etcetera. Of this, Reoviridae is the most diverse virus. These reoviruses can cause varied number of diseases including respiratory and enteric diseases, Colorado tick fever, gastroenteritis and diarrhoea. The chapter serves as a source to understand the major categories related to the varied types of RNA viruses.

Chapter 6- Since cell DNA takes place during the S phase only, the viruses that depend on the DNA polymerase of the host cell needs to wait to enter the S phase to regulate cell cycle. Large DNA viruses possess their own intermediate size DNA and DNA machinery. The aspects elucidated in this sec-tion are of vital importance, and provide a better understanding of DNA viruses.

I would like to make a special mention of my publisher who considered me worthy of this opportunity and also supported me throughout the process. I would also like to thank the editing team at the back-end who extended their help whenever required.

Editor

1

Understanding Virology

Virus is an agent that is harmful in nature and can replicate and multiply when inside an organism. Virology is the study of these viruses, their structure, evolution and their influence on the host body. According to shapes, viruses can be classified into icosahedral and helical symmetries. The chapter on virus offers an insightful focus, keeping in mind the complex subject matter.

Virology

Virology is the study of viruses – submicroscopic, parasitic particles of genetic material contained in a protein coat – and virus-like agents. It focuses on the following aspects of viruses: their structure, classification and evolution, their ways to infect and exploit host cells for reproduction, their interaction with host organism physiology and immunity, the diseases they cause, the techniques to isolate and culture them, and their use in research and therapy. Virology is considered to be a subfield of microbiology or of medicine.

Virus Structure and Classification

A major branch of virology is virus classification. Viruses can be classified according to the host cell they infect: animal viruses, plant viruses, fungal viruses, and bacteriophages (viruses infecting bacterium, which include the most complex viruses). Another classification uses the geometrical shape of their capsid (often a helix or an icosahedron) or the virus's structure (e.g. presence or absence of a lipid envelope). Viruses range in size from about 30 nm to about 450 nm, which means that most of them cannot be seen with light microscopes. The shape and structure of viruses has been studied by electron microscopy, NMR spectroscopy, and X-ray crystallography.

The most useful and most widely used classification system distinguishes viruses according to the type of nucleic acid they use as genetic material and the viral replication method they employ to coax host cells into producing more viruses:

- DNA viruses (divided into double-stranded DNA viruses and single-stranded DNA viruses),
- RNA viruses (divided into positive-sense single-stranded RNA viruses, negative-sense single-stranded RNA viruses and the much less common double-stranded RNA viruses),
- reverse transcribing viruses (double-stranded reverse-transcribing DNA viruses and single-stranded reverse-transcribing RNA viruses including retroviruses).

The latest report by the International Committee on Taxonomy of Viruses (2005) lists 5450 viruses, organized in over 2,000 species, 287 genera, 73 families and 3 orders.

Virologists also study *subviral particles*, infectious entities notably smaller and simpler than viruses:

- viroids (naked circular RNA molecules infecting plants),
- satellites (nucleic acid molecules with or without a capsid that require a helper virus for infection and reproduction), and
- prions (proteins that can exist in a pathological conformation that induces other prion molecules to assume that same conformation).

Taxa in virology are not necessarily monophyletic, as the evolutionary relationships of the various virus groups remain unclear. Three hypotheses regarding their origin exist:

1. Viruses arose from non-living matter, separately from yet in parallel to cells, perhaps in the form of self-replicating RNA ribozymes similar to viroids.
2. Viruses arose by genome reduction from earlier, more competent cellular life forms that became parasites to host cells and subsequently lost most of their functionality; examples of such tiny parasitic prokaryotes are Mycoplasma and Nanoarchaea.
3. Viruses arose from mobile genetic elements of cells (such as transposons, retrotransposons or plasmids) that became encapsulated in protein capsids, acquired the ability to "break free" from the host cell and infect other cells.

Of particular interest here is mimivirus, a giant virus that infects amoebae and encodes much of the molecular machinery traditionally associated with bacteria. Is it a simplified version of a parasitic prokaryote, or did it originate as a simpler virus that acquired genes from its host?

The evolution of viruses, which often occurs in concert with the evolution of their hosts, is studied in the field of viral evolution.

While viruses reproduce and evolve, they do not engage in metabolism, do not move, and depend on a host cell for reproduction. The often-debated question of whether they are alive or not is a matter of definition that does not affect the biological reality of viruses.

Viral Diseases and Host Defenses

One main motivation for the study of viruses is the fact that they cause many important infectious diseases, among them the common cold, influenza, rabies, measles, many forms of diarrhea, hepatitis, Dengue fever, yellow fever, polio, smallpox and AIDS. Herpes simplex causes cold sores and genital herpes and is under investigation as a possible factor in Alzheimer's.

Some viruses, known as oncoviruses, contribute to the development of certain forms of cancer. The best studied example is the association between Human papillomavirus and cervical cancer: almost all cases of cervical cancer are caused by certain strains of this sexually transmitted virus. Another example is the association of infection with hepatitis B and hepatitis C viruses and liver cancer.

Some subviral particles also cause disease: the transmissible spongiform encephalopathies, which include Kuru, Creutzfeldt–Jakob disease and bovine spongiform encephalopathy ("mad cow disease"), are caused by prions, hepatitis D is due to a satellite virus.

The study of the manner in which viruses cause disease is viral pathogenesis. The degree to which a virus causes disease is its virulence.

When the immune system of a vertebrate encounters a virus, it may produce specific antibodies which bind to the virus and neutralize its infectivity or mark it for destruction. Antibody presence in blood serum is often used to determine whether a person has been exposed to a given virus in the past, with tests such as ELISA. Vaccinations protect against viral diseases, in part, by eliciting the production of antibodies. Monoclonal antibodies, specific to the virus, are also used for detection, as in fluorescence microscopy.

A second defense of vertebrates against viruses, cell-mediated immunity, involves immune cells known as T cells: the body's cells constantly display short fragments of their proteins on the cell's surface, and if a T cell recognizes a suspicious viral fragment there, the host cell is destroyed and the virus-specific T-cells proliferate. This mechanism is jump-started by certain vaccinations.

RNA interference, an important cellular mechanism found in plants, animals and many other eukaryotes, most likely evolved as a defense against viruses. An elaborate machinery of interacting enzymes detects double-stranded RNA molecules (which occur as part of the life cycle of many viruses) and then proceeds to destroy all single-stranded versions of those detected RNA molecules.

Every lethal viral disease presents a paradox: killing its host is obviously of no benefit to the virus, so how and why did it evolve to do so? Today it is believed that most viruses are relatively benign in their natural hosts; some viral infection might even be beneficial to the host. The lethal viral diseases are believed to have resulted from an "accidental" jump of the virus from a species in which it is benign to a new one that is not accustomed to it. For example, viruses that cause serious influenza in humans probably have pigs or birds as their natural host, and HIV is thought to derive from the benign non-human primate virus SIV.

While it has been possible to prevent (certain) viral diseases by vaccination for a long time, the development of antiviral drugs to *treat* viral diseases is a comparatively recent development. The first such drug was interferon, a substance that is naturally produced when an infection is detected and stimulates other parts of the immune system.

Molecular Biology Research and Viral Therapy

Bacteriophages, the viruses which infect bacteria, can be relatively easily grown as viral plaques on bacterial cultures. Bacteriophages occasionally move genetic material from one bacterial cell to another in a process known as transduction, and this horizontal gene transfer is one reason why they served as a major research tool in the early development of molecular biology. The genetic code, the function of ribozymes, the first recombinant DNA and early genetic libraries were all arrived at using bacteriophages. Certain genetic elements derived from viruses, such as highly effective promoters, are commonly used in molecular biology research today.

Growing animal viruses outside of the living host animal is more difficult. Classically, fertilized chicken eggs have often been used, but cell cultures are increasingly employed for this purpose today.

Since some viruses that infect eukaryotes need to transport their genetic material into the host cell's nucleus, they are attractive tools for introducing new genes into the host (known as transformation or transfection). Modified retroviruses are often used for this purpose, as they integrate their genes into the host's chromosomes.

This approach of using viruses as gene vectors is being pursued in the gene therapy of genetic diseases. An obvious problem to be overcome in viral gene therapy is the rejection of the transforming virus by the immune system.

Phage therapy, the use of bacteriophages to combat bacterial diseases, was a popular research topic before the advent of antibiotics and has recently seen renewed interest.

Oncolytic viruses are viruses that preferably infect cancer cells. While early efforts to employ these viruses in the therapy of cancer failed, there have been reports in 2005 and 2006 of encouraging preliminary results.

Other uses of Viruses

A new application of genetically engineered viruses in nanotechnology was recently described.

History of Virology

The word *virus* appeared in 1599 and originally meant "venom".

A very early form of vaccination known as variolation was developed several thousand years ago in China. It involved the application of materials from smallpox sufferers in order to immunize others. In 1717 Lady Mary Wortley Montagu observed the practice in Istanbul and attempted to popularize it in Britain, but encountered considerable resistance. In 1796 Edward Jenner developed a much safer method, using cowpox to successfully immunize a young boy against smallpox, and this practice was widely adopted. Vaccinations against other viral diseases followed, including the successful rabies vaccination by Louis Pasteur in 1886. The nature of viruses however was not clear to these researchers.

Dmitri Ivanovsky

Martinus Beijerinck

In 1892, the Russian biologist Dmitry Ivanovsky used a Chamberland filter to try to isolate the bacteria that caused tobacco mosaic disease. His experiments showed that crushed leaf extracts from infected tobacco plants remained infectious after filtration. Ivanovsky reported a minuscule infectious agent or toxin, capable of passing the filter, may be being produced by a bacterium.

In 1898 Martinus Beijerinck repeated Ivanovski's work but went further and passed the "filterable agent" from plant to plant, found the action undiminished, and concluded it infectious—replicating in the host—and thus not a mere toxin. He called it *contagium vivum fluidum*. The question of whether the agent was a "living fluid" or a particle was however still open.

In 1903 it was suggested for the first time that transduction by viruses might cause cancer. In 1908 Bang and Ellerman showed that a filterable virus could transmit chicken leukemia, data largely ignored till the 1930s when leukemia became regarded as cancerous. In 1911 Peyton Rous reported the transmission of chicken sarcoma, a solid tumor, with a virus, and thus Rous became "father of tumor virology". The virus was later called Rous sarcoma virus 1 and understood to be a retrovirus. Several other cancer-causing retroviruses have since been described.

The existence of viruses that infect bacteria (bacteriophages) was first recognized by Frederick Twort in 1911, and, independently, by Félix d'Herelle in 1917. As bacteria could be grown easily in culture, this led to an explosion of virology research.

The cause of the devastating Spanish flu pandemic of 1918 was initially unclear. In late 1918, French scientists showed that a "filter-passing virus" could transmit the disease to people and animals, fulfilling Koch's postulates.

In 1926 it was shown that scarlet fever is caused by a bacterium that is infected by a certain bacteriophage.

While plant viruses and bacteriophages can be grown comparatively easily, animal viruses normally require a living host animal, which complicates their study immensely. In 1931 it was shown that influenza virus could be grown in fertilized chicken eggs, a method that is still used today to produce vaccines. In 1937, Max Theiler managed to grow the yellow fever virus in chicken eggs and produced a vaccine from an attenuated virus strain; this vaccine saved millions of lives and is still being used today.

Max Delbrück, an important investigator in the area of bacteriophages, described the basic "life cycle" of a virus in 1937: rather than "growing", a virus particle is assembled from its constituent pieces in one step; eventually it leaves the host cell to infect other cells. The Hershey–Chase experiment in 1952 showed that only DNA and not protein enters a bacterial cell upon infection with bacteriophage T2. Transduction of bacteria by bacteriophages was first described in the same year.

In 1949 John F. Enders, Thomas Weller and Frederick Robbins reported growth of poliovirus in cultured human embryonal cells, the first significant example of an animal virus grown outside of animals or chicken eggs. This work aided Jonas Salk in deriving a polio vaccine from deactivated polio viruses; this vaccine was shown to be effective in 1955.

The first virus that could be crystalized and whose structure could therefore be elucidated in detail was tobacco mosaic virus (TMV), the virus that had been studied earlier by Ivanovski and Beijerink. In 1935, Wendell Stanley achieved its crystallization for electron microscopy and showed that it remains active even after crystallization. Clear X-ray diffraction pictures of the crystallized virus were obtained by Bernal and Fankuchen in 1941. Based on such pictures, Rosalind Franklin proposed the full structure of the tobacco mosaic virus in 1955. Also in 1955, Heinz Fraenkel-Conrat and Robley Williams showed that purified TMV RNA and its capsid (coat) protein can self-assemble into functional virions, suggesting that this assembly mechanism is also used within the host cell, as Delbrück had proposed earlier.

From the 1950s-60's, Chester M. Southam, a leading virologist and cancer researcher, injected cancer patients, healthy individuals, and prison inmates from the Ohio Penitentiary with HeLa cancer cells in order to observe if cancer could be transmitted. Additionally, in hopes of creating a vaccine for cancer, he observed if the subjects could become immune to cancer by developing an acquired immune response. This experiment was highly controversial, as the cancer patient subjects were unaware that they were being injected with cancer cells.

In 1963, the Hepatitis B virus was discovered by Baruch Blumberg who went on to develop a hepatitis B vaccine.

In 1965, Howard Temin described the first retrovirus: a virus whose RNA genome was reverse transcribed into complementary DNA (cDNA), then integrated into the host's genome and expressed from that template. The viral enzyme reverse transcriptase, which along with integrase is a distinguishing trait of retroviruses, was first described in 1970, independently by Howard Temin and David Baltimore. The first retrovirus infecting humans was identified by Robert Gallo in 1974. Later it was found that reverse transcriptase is not specific to retroviruses; retrotransposons which code for reverse transcriptase are abundant in the genomes of all eukaryotes. About 10-40% of the human genome derives from such retrotransposons.

In 1975 the functioning of oncoviruses was clarified considerably. Until that time, it was thought that these viruses carried certain genes called oncogenes which, when inserted into the host's genome, would cause cancer. Michael Bishop and Harold Varmus showed that the oncogene of Rous sarcoma virus is in fact not specific to the virus but is contained in the genome of healthy animals of many species. The oncovirus can switch this pre-existing benign proto-oncogene on, turning it into a true oncogene that causes cancer.

1976 saw the first recorded outbreak of Ebola virus disease, a highly lethal virally transmitted disease.

In 1977, Frederick Sanger achieved the first complete sequencing of the genome of any organism, the bacteriophage Phi X 174. In the same year, Richard Roberts and Phillip Sharp independently showed that the genes of adenovirus contain introns and therefore require gene splicing. It was later realized that almost all genes of eukaryotes have introns as well.

A worldwide vaccination campaign led by the UN World Health Organization resulted in the eradication of smallpox in 1979.

In 1982, Stanley Prusiner discovered prions and showed that they cause scrapie.

The first cases of AIDS were reported in 1981, and HIV, the retrovirus causing it, was identified in 1983 by Luc Montagnier, Françoise Barré-Sinoussi and Robert Gallo. Tests detecting HIV infection by detecting the presence of HIV antibody were developed. Subsequent tremendous research efforts turned HIV into the best studied virus. Human Herpes Virus 8, the cause of Kaposi's sarcoma which is often seen in AIDS patients, was identified in 1994. Several antiretroviral drugs were developed in the late 1990s, decreasing AIDS mortality dramatically in developed countries.

The Hepatitis C virus was identified using novel molecular cloning techniques in 1987, leading to screening tests that dramatically reduced the incidence of post-transfusion hepatitis.

The first attempts at gene therapy involving viral vectors began in the early 1980s, when retroviruses were developed that could insert a foreign gene into the host's genome. They contained the foreign gene but did not contain the viral genome and therefore could not reproduce. Tests in mice were followed by tests in humans, beginning in 1989. The first human studies attempted to correct the genetic disease severe combined immunodeficiency (SCID), but clinical success was limited. In the period from 1990 to 1995, gene therapy was tried on several other diseases and with different viral vectors, but it became clear that the initially high expectations were overstated. In 1999 a further setback occurred when 18-year-old Jesse Gelsinger died in a gene therapy trial. He suffered a severe immune response after having received an adenovirus vector. Success in the gene therapy of two cases of X-linked SCID was reported in 2000.

In 2002 it was reported that poliovirus had been synthetically assembled in the laboratory, representing the first synthetic organism. Assembling the 7741-base genome from scratch, starting with the virus's published RNA sequence, took about two years. In 2003 a faster method was shown to assemble the 5386-base genome of the bacteriophage Phi X 174 in 2 weeks.

The giant mimivirus, in some sense an intermediate between tiny prokaryotes and ordinary viruses, was described in 2003 and sequenced in 2004.

The strain of Influenza A virus subtype H1N1 that killed up to 50 million people during the Spanish flu pandemic in 1918 was reconstructed in 2005. Sequence information was pieced together from preserved tissue samples of flu victims; viable virus was then synthesized from this sequence. The 2009 flu pandemic involved another strain of Influenza A H1N1, commonly known as "swine flu".

By 1985, Harald zur Hausen had shown that two strains of Human papillomavirus (HPV) cause most cases of cervical cancer. Two vaccines protecting against these strains were released in 2006.

In 2006 and 2007 it was reported that introducing a small number of specific transcription factor genes into normal skin cells of mice or humans can turn these cells into pluripotent stem cells, known as induced pluripotent stem cells. The technique uses modified retroviruses to transform the cells; this is a potential problem for human therapy since these viruses integrate their genes at a random location in the host's genome, which can interrupt other genes and potentially causes cancer.

In 2008, Sputnik virophage was described, the first known *virophage*: it uses the machinery of a helper virus to reproduce and inhibits reproduction of that helper virus. Sputnik reproduces in amoeba infected by mamavirus, a relative of the mimivirus mentioned above and the largest known virus to date.

An endogenous retrovirus (ERV) is a retrovirus whose genome has been permanently incorporated into the germ-line genome of some organism and that is therefore copied with each reproduction of that organism. It is estimated that about 9 percent of the human genome have their origin in ERVs. In 2015 it was shown that proteins from an ERV are actively expressed in 3-day-old human embryos and appear to play a role in embryonal development and protect embryos from infection by other viruses.

Virus

A virus is a small infectious agent that replicates only inside the living cells of other organisms. Viruses can infect all types of life forms, from animals and plants to microorganisms, including bacteria and archaea.

Since Dmitri Ivanovsky's 1892 article describing a non-bacterial pathogen infecting tobacco plants, and the discovery of the tobacco mosaic virus by Martinus Beijerinck in 1898, about 5,000 virus species have been described in detail, although there are millions of types. Viruses are found in almost every ecosystem on Earth and are the most abundant type of biological entity. The study of viruses is known as virology, a sub-speciality of microbiology.

While not inside an infected cell or in the process of infecting a cell, viruses exist in the form of independent particles. These viral particles, also known as virions, consist of two or three parts: (i) the genetic material made from either DNA or RNA, long molecules that carry genetic information; (ii) a protein coat, called the capsid, which surrounds and protects the genetic material; and in some cases (iii) an envelope of lipids that surrounds the protein coat when they are outside a cell. The shapes of these virus particles range from simple helical and icosahedral forms for some virus species to more complex structures for others. Most virus species have virions that are too small to be seen with an optical microscope. The average virion is about one one-hundredth the size of the average bacterium.

The origins of viruses in the evolutionary history of life are unclear: some may have evolved from plasmids—pieces of DNA that can move between cells—while others may have evolved from bacteria. In evolution, viruses are an important means of horizontal gene transfer, which increases genetic diversity. Viruses are considered by some to be a life form, because they carry genetic material, reproduce, and evolve through natural selection. However they lack key characteristics (such as cell structure) that are generally considered necessary to count as life. Because they possess some but not all such qualities, viruses have been described as "organisms at the edge of life", and as replicators.

Viruses spread in many ways; viruses in plants are often transmitted from plant to plant by insects that feed on plant sap, such as aphids; viruses in animals can be carried by blood-sucking insects. These disease-bearing organisms are known as vectors. Influenza viruses are spread by coughing and sneezing. Norovirus and rotavirus, common causes of viral gastroenteritis, are transmitted by the faecal–oral route and are passed from person to person by contact, entering the body in food or water. HIV is one of several viruses transmitted through sexual contact and by exposure to infected blood. The range of host cells that a virus can infect is called its "host range". This can be narrow, meaning a virus is capable of infecting few species, or broad, meaning it is capable of infecting many.

Viral infections in animals provoke an immune response that usually eliminates the infecting virus. Immune responses can also be produced by vaccines, which confer an artificially acquired immunity to the specific viral infection. However, some viruses including those that cause AIDS and viral hepatitis evade these immune responses and result in chronic infections. Antibiotics have no effect on viruses, but several antiviral drugs have been developed.

Etymology

The word is from the Latin neuter *vīrus* referring to poison and other noxious liquids, from 'the same Indo-European base as Sanskrit *visa* poison, Avestan *vīša* poison, ancient Greek poison', first attested in English in 1398 in John Trevisa's translation of Bartholomeus Anglicus's *De Proprietatibus Rerum. Virulent*, from Latin *virulentus* (poisonous), dates to c. 1400. A meaning of "agent that causes infectious disease" is first recorded in 1728, before the discovery of viruses by Dmitri Ivanovsky in 1892. The English plural is *viruses* (sometimes also *viri* or *vira*), whereas the Latin word is a mass noun, which has no classically attested plural (however in Neo-Latin *vīra* is used). The adjective *viral* dates to 1948. The term *virion* (plural *virions*), which dates from 1959, is also used to refer to a single, stable infective viral particle that is released from the cell and is fully capable of infecting other cells of the same type.

History

Louis Pasteur was unable to find a causative agent for rabies and speculated about a pathogen too small to be detected using a microscope. In 1884, the French microbiologist Charles Chamberland invented a filter (known today as the Chamberland filter or the Pasteur-Chamberland filter) with pores smaller than bacteria. Thus, he could pass a solution containing bacteria through the filter and completely remove them from the solution. In 1892, the Russian biologist Dmitri Ivanovsky used this filter to study what is now known as the tobacco mosaic virus. His experiments showed that crushed leaf extracts from infected tobacco plants remain infectious after filtration. Ivanovsky suggested the infection might be caused by a toxin produced by bacteria, but did not pursue the idea. At the time it was thought that all infectious agents could be retained by filters and grown on a nutrient medium – this was part of the germ theory of disease. In 1898, the Dutch microbiologist Martinus Beijerinck repeated the experiments and became convinced that the filtered solution contained a new form of infectious agent. He observed that the agent multiplied only in cells that were dividing, but as his experiments did not show that it was made of particles, he called it a *contagium vivum fluidum* (soluble living germ) and re-introduced the word *virus*. Beijerinck maintained that viruses were liquid in nature, a theory later discredited by Wendell Stanley, who proved they were particulate. In the same year Friedrich Loeffler and Paul Frosch passed the first animal virus – agent of foot-and-mouth disease (aphthovirus) – through a similar filter.

Martinus Beijerinck in his laboratory in 1921

In the early 20th century, the English bacteriologist Frederick Twort discovered a group of viruses that infect bacteria, now called bacteriophages (or commonly *phages*), and the French-Canadian microbiologist Félix d'Herelle described viruses that, when added to bacteria on an agar plate, would produce areas of dead bacteria. He accurately diluted a suspension of these viruses and discovered that the highest dilutions (lowest virus concentrations), rather than killing all the bacteria, formed discrete areas of dead organisms. Counting these areas and multiplying by the dilution factor allowed him to calculate the number of viruses in the original suspension. Phages were heralded as a potential treatment for diseases such as typhoid and cholera, but their promise was forgotten with the development of penicillin. The study of phages provided insights into the switching on and off of genes, and a useful mechanism for introducing foreign genes into bacteria.

By the end of the 19th century, viruses were defined in terms of their infectivity, their ability to be filtered, and their requirement for living hosts. Viruses had been grown only in plants and animals. In 1906, Ross Granville Harrison invented a method for growing tissue in lymph, and, in 1913, E. Steinhardt, C. Israeli, and R. A. Lambert used this method to grow vaccinia virus in fragments of guinea pig corneal tissue. In 1928, H. B. Maitland and M. C. Maitland grew vaccinia virus in suspensions of minced hens' kidneys. Their method was not widely adopted until the 1950s, when poliovirus was grown on a large scale for vaccine production.

Another breakthrough came in 1931, when the American pathologist Ernest William Goodpasture and Alice Miles Woodruff grew influenza and several other viruses in fertilised chickens' eggs. In 1949, John Franklin Enders, Thomas Weller, and Frederick Robbins grew polio virus in cultured human embryo cells, the first virus to be grown without using solid animal tissue or eggs. This work enabled Jonas Salk to make an effective polio vaccine.

The first images of viruses were obtained upon the invention of electron microscopy in 1931 by the German engineers Ernst Ruska and Max Knoll. In 1935, American biochemist and virologist Wendell Meredith Stanley examined the tobacco mosaic virus and found it was mostly made of protein. A short time later, this virus was separated into protein and RNA parts. The tobacco mosaic virus was the first to be crystallised and its structure could therefore be elucidated in detail. The first X-ray diffraction pictures of the crystallised virus were obtained by Bernal and Fankuchen in 1941. On the basis of her pictures, Rosalind Franklin discovered the full structure of the virus in 1955. In the same year, Heinz Fraenkel-Conrat and Robley Williams showed that purified tobacco mosaic virus RNA and its protein coat can assemble by themselves to form functional viruses, suggesting that this simple mechanism was probably the means through which viruses were created within their host cells.

The second half of the 20th century was the golden age of virus discovery and most of the over 2,000 recognised species of animal, plant, and bacterial viruses were discovered during these years. In 1957, equine arterivirus and the cause of Bovine virus diarrhoea (a pestivirus) were discovered. In 1963, the hepatitis B virus was discovered by Baruch Blumberg, and in 1965, Howard Temin described the first retrovirus. Reverse transcriptase, the enzyme that retroviruses use to make DNA copies of their RNA, was first described in 1970, independently by Howard Martin Temin and David Baltimore. In 1983 Luc Montagnier's team at the Pasteur Institute in France, first isolated the retrovirus now called HIV. In 1989 Michael Houghton's team at Chiron Corporation discovered Hepatitis C.

Viruses are found wherever there is life and have probably existed since living cells first evolved. The origin of viruses is unclear because they do not form fossils, so molecular techniques have been used to compare the DNA or RNA of viruses and are a useful means of investigating how they arose. In addition, viral genetic material may occasionally integrate into the germline of the host organisms, by which they can be passed on vertically to the offspring of the host for many generations. This provides an invaluable source of information for paleovirologists to trace back ancient viruses that have existed up to millions of years ago. There are three main hypotheses that aim to explain the origins of viruses:

Regressive Hypothesis

Viruses may have once been small cells that parasitised larger cells. Over time, genes not required by their parasitism were lost. The bacteria rickettsia and chlamydia are living cells that, like viruses, can reproduce only inside host cells. They lend support to this hypothesis, as their dependence on parasitism is likely to have caused the loss of genes that enabled them to survive outside a cell. This is also called the *degeneracy hypothesis*, or *reduction hypothesis*.

Cellular Origin Hypothesis

Some viruses may have evolved from bits of DNA or RNA that "escaped" from the genes of a larger organism. The escaped DNA could have come from plasmids (pieces of naked DNA that can move *between* cells) or transposons (molecules of DNA that replicate and move around to different positions *within* the genes of the cell). Once called "jumping genes", transposons are examples of mobile genetic elements and could be the origin of some viruses. They were discovered in maize by Barbara McClintock in 1950. This is sometimes called the *vagrancy hypothesis*, or the *escape hypothesis*.

Co-evolution Hypothesis

This is also called the *virus-first hypothesis* and proposes that viruses may have evolved from complex molecules of protein and nucleic acid at the same time as cells first appeared on Earth and would have been dependent on cellular life for billions of years. Viroids are molecules of RNA that are not classified as viruses because they lack a protein coat. However, they have characteristics that are common to several viruses and are often called subviral agents. Viroids are important pathogens of plants. They do not code for proteins but interact with the host cell and use the host machinery for their replication. The hepatitis delta virus of humans has an RNA genome similar to viroids but has a protein coat derived from hepatitis B virus and cannot produce one of its own. It is, therefore, a defective virus. Although hepatitis delta virus genome may replicate independently once inside a host cell, it requires the help of hepatitis B virus to provide a protein coat so that it can be transmitted to new cells. In similar manner, the sputnik virophage is dependent on mimivirus, which infects the protozoan *Acanthamoeba castellanii*. These viruses, which are dependent on the presence of other virus species in the host cell, are called *satellites* and may represent evolutionary intermediates of viroids and viruses.

In the past, there were problems with all of these hypotheses: the regressive hypothesis did not explain why even the smallest of cellular parasites do not resemble viruses in any way. The escape hypothesis did not explain the complex capsids and other structures on virus particles. The virus-first hypothesis contravened the definition of viruses in that they require host cells. Viruses are now recognised as ancient and as having origins that pre-date the divergence of life into the three domains. This discovery has led modern virologists to reconsider and re-evaluate these three classical hypotheses.

The evidence for an ancestral world of RNA cells and computer analysis of viral and host DNA sequences are giving a better understanding of the evolutionary relationships between different viruses and may help identify the ancestors of modern viruses. To date, such analyses have not proved which of these hypotheses is correct. However, it seems unlikely that all currently known viruses have a common ancestor, and viruses have probably arisen numerous times in the past by one or more mechanisms.

Prions are infectious protein molecules that do not contain DNA or RNA. They can cause infections such as scrapie in sheep, bovine spongiform encephalopathy ("mad cow" disease) in cattle, and chronic wasting disease in deer; in humans, prionic diseases include Kuru, Creutzfeldt–Jakob disease, and Gerstmann–Sträussler–Scheinker syndrome. Although prions are fundamentally different from viruses and viroids, their discovery gives credence to the theory that viruses could have evolved from self-replicating molecules.

Microbiology

Life Properties

Opinions differ on whether viruses are a form of life, or organic structures that interact with living organisms. They have been described as "organisms at the edge of life", since they resemble organisms in that they possess genes, evolve by natural selection, and reproduce by creating multiple copies of themselves through self-assembly. Although they have genes, they do not have a cellular structure, which is often seen as the basic unit of life. Viruses do not have their own metabolism, and require a host cell to make new products. They therefore cannot naturally reproduce outside a host cell – although bacterial species such as rickettsia and chlamydia are considered living organisms despite the same limitation. Accepted forms of life use cell division to reproduce, whereas viruses spontaneously assemble within cells. They differ from autonomous growth of crystals as they inherit genetic mutations while being subject to natural selection. Virus self-assembly within host cells has implications for the study of the origin of life, as it lends further credence to the hypothesis that life could have started as self-assembling organic molecules.

Structure

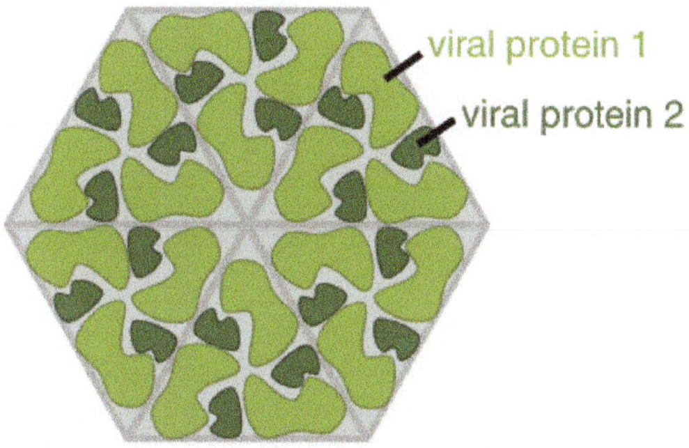

Diagram of how a virus capsid can be constructed using multiple copies of just two protein molecules

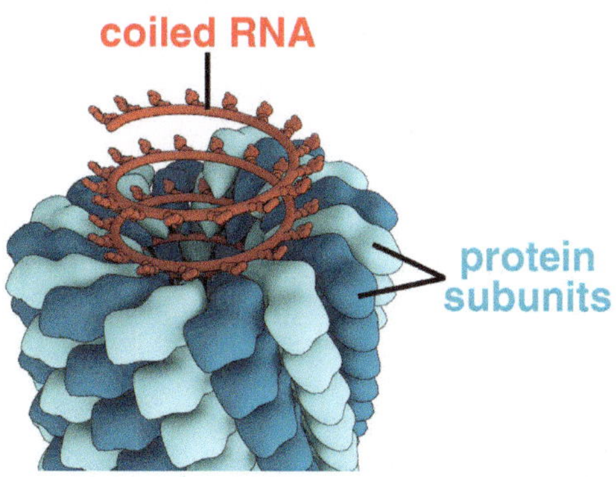

Structure of tobacco mosaic virus: RNA coiled in a helix of repeating protein sub-units

Structure of icosahedral adenovirus. Electron micrograph of with a cartoon to show shape

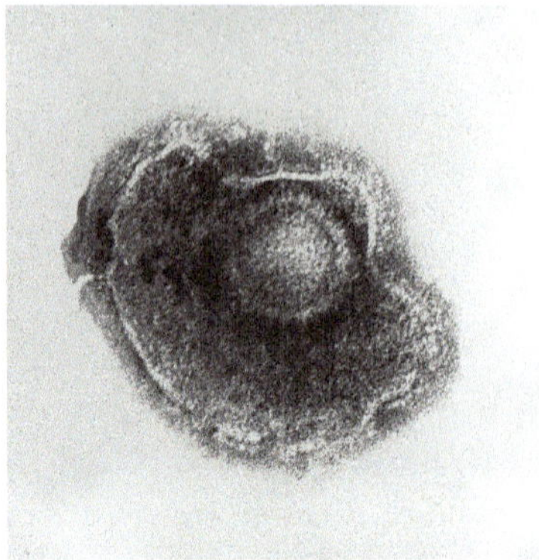

Structure of chickenpox virus. They have a lipid envelope

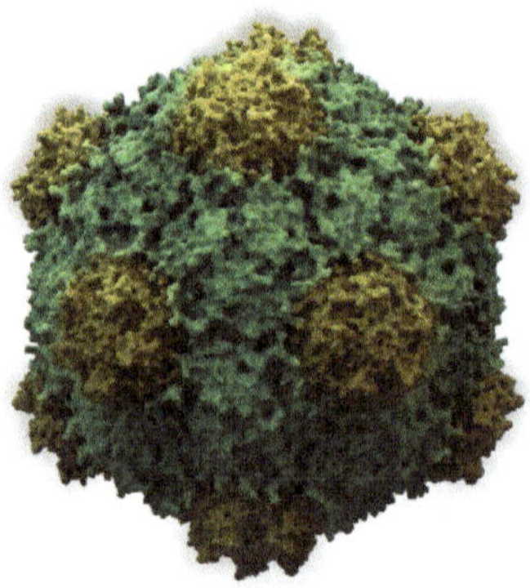

Structure of an icosahedral cowpea mosaic virus

Viruses display a wide diversity of shapes and sizes, called *morphologies*. In general, viruses are much smaller than bacteria. Most viruses that have been studied have a diameter between 20 and 300 nanometres. Some filoviruses have a total length of up to 1400 nm; their diameters are only about 80 nm. Most viruses cannot be seen with an optical microscope so scanning and transmission electron microscopes are used to visualise virions. To increase the contrast between viruses and the background, electron-dense "stains" are used. These are solutions of salts of heavy metals, such as tungsten, that scatter the electrons from regions covered with the stain. When virions are coated with stain (positive staining), fine detail is obscured. Negative staining overcomes this problem by staining the background only.

A complete virus particle, known as a virion, consists of nucleic acid surrounded by a protective coat of protein called a capsid. These are formed from identical protein subunits called capsomeres. Viruses can have a lipid "envelope" derived from the host cell membrane. The capsid is made from proteins encoded by the viral genome and its shape serves as the basis for morphological distinction. Virally coded protein subunits will self-assemble to form a capsid, in general requiring the presence of the virus genome. Complex viruses code for proteins that assist in the construction of their capsid. Proteins associated with nucleic acid are known as nucleoproteins, and the association of viral capsid proteins with viral nucleic acid is called a nucleocapsid. The capsid and entire virus structure can be mechanically (physically) probed through atomic force microscopy. In general, there are four main morphological virus types:

Helical

These viruses are composed of a single type of capsomere stacked around a central axis to form a helical structure, which may have a central cavity, or tube. This arrangement results in rod-shaped or filamentous virions: These can be short and highly rigid, or long and very flexible. The genetic material, in general, single-stranded RNA, but ssDNA in some cases, is bound into the protein helix by interactions between the negatively charged nucleic acid and positive charges on the protein. Overall, the length of a helical capsid is related to the length of the nucleic acid contained within it and the diameter is dependent on the size and arrangement of capsomeres. The well-studied tobacco mosaic virus is an example of a helical virus.

Icosahedral

Most animal viruses are icosahedral or near-spherical with chiral icosahedral symmetry. A regular icosahedron is the optimum way of forming a closed shell from identical sub-units. The minimum number of identical capsomeres required is twelve, each composed of five identical sub-units. Many viruses, such as rotavirus, have more than twelve capsomers and appear spherical but they retain this symmetry. Capsomeres at the apices are surrounded by five other capsomeres and are called pentons. Capsomeres on the triangular faces are surrounded by six others and are called hexons. Hexons are in essence flat and pentons, which form the 12 vertices, are curved. The same protein may act as the subunit of both the pentamers and hexamers or they may be composed of different proteins.

Prolate

This is an icosahedron elongated along the fivefold axis and is a common arrangement of the heads of bacteriophages. This structure is composed of a cylinder with a cap at either end.

Envelope

Some species of virus envelop themselves in a modified form of one of the cell membranes, either the outer membrane surrounding an infected host cell or internal membranes such as nuclear membrane or endoplasmic reticulum, thus gaining an outer lipid bilayer known as a viral envelope. This membrane is studded with proteins coded for by the viral genome and host genome; the lipid membrane itself and any carbohydrates present originate entirely from the host. The influenza virus and HIV use this strategy. Most enveloped viruses are dependent on the envelope for their infectivity.

Complex

These viruses possess a capsid that is neither purely helical nor purely icosahedral, and that may possess extra structures such as protein tails or a complex outer wall. Some bacteriophages, such as Enterobacteria phage T4, have a complex structure consisting of an icosahedral head bound to a helical tail, which may have a hexagonal base plate with protruding protein tail fibres. This tail structure acts like a molecular syringe, attaching to the bacterial host and then injecting the viral genome into the cell.

The poxviruses are large, complex viruses that have an unusual morphology. The viral genome is associated with proteins within a central disc structure known as a nucleoid. The nucleoid is surrounded by a membrane and two lateral bodies of unknown function. The virus has an outer envelope with a thick layer of protein studded over its surface. The whole virion is slightly pleiomorphic, ranging from ovoid to brick shape. Mimivirus is one of the largest characterised viruses, with a capsid diameter of 400 nm. Protein filaments measuring 100 nm project from the surface. The capsid appears hexagonal under an electron microscope, therefore the capsid is probably icosahedral. In 2011, researchers discovered the largest then known virus in samples of water collected from the ocean floor off the coast of Las Cruces, Chile. Provisionally named *Megavirus chilensis*, it can be seen with a basic optical microscope. In 2013, the Pandoravirus genus was discovered in Chile and Australia, and has genomes about twice as large as Megavirus and Mimivirus.

Some viruses that infect Archaea have complex structures that are unrelated to any other form of virus, with a wide variety of unusual shapes, ranging from spindle-shaped structures, to viruses that resemble hooked rods, teardrops or even bottles. Other archaeal viruses resemble the tailed bacteriophages, and can have multiple tail structures.

Genome

Genomic diversity among viruses	
Property	Parameters
Nucleic acid	• DNA • RNA • Both DNA and RNA (at different stages in the life cycle)
Shape	• Linear • Circular • Segmented
Strandedness	• Single-stranded • Double-stranded • Double-stranded with regions of single-strandedness
Sense	• Positive sense (+) • Negative sense (−) • Ambisense (+/−)

An enormous variety of genomic structures can be seen among viral species; as a group, they contain more structural genomic diversity than plants, animals, archaea, or bacteria. There are millions of different types of viruses, although only about 5,000 types have been described in detail. As of September 2015, the NCBI Virus genome database has more than 75,000 complete genome sequences. but there are doubtlessly many more to be discovered.

A virus has either a DNA or an RNA genome and is called a DNA virus or an RNA virus, respectively. The vast majority of viruses have RNA genomes. Plant viruses tend to have single-stranded RNA genomes and bacteriophages tend to have double-stranded DNA genomes.

Viral genomes are *circular*, as in the polyomaviruses, or *linear*, as in the adenoviruses. The type of nucleic acid is irrelevant to the shape of the genome. Among RNA viruses and certain DNA viruses, the genome is often divided up into separate parts, in which case it is called segmented. For RNA viruses, each segment often codes for only one protein and they are usually found together in one capsid. However, all segments are not required to be in the same virion for the virus to be infectious, as demonstrated by brome mosaic virus and several other plant viruses.

A viral genome, irrespective of nucleic acid type, is almost always either *single-stranded* or *double-stranded*. Single-stranded genomes consist of an unpaired nucleic acid, analogous to one-half of a ladder split down the middle. Double-stranded genomes consist of two complementary paired nucleic acids, analogous to a ladder. The virus particles of some virus families, such as those belonging to the *Hepadnaviridae*, contain a genome that is partially double-stranded and partially single-stranded.

For most viruses with RNA genomes and some with single-stranded DNA genomes, the single strands are said to be either positive-sense (called the *plus-strand*) or negative-sense (called the *minus-strand*), depending on if they are complementary to the viral messenger RNA (mRNA). Positive-sense viral RNA is in the same sense as viral mRNA and thus at least a part of it can be immediately translated by the host cell. Negative-sense viral RNA is complementary to mRNA and thus must be converted to positive-sense RNA by an RNA-dependent RNA polymerase before translation. DNA nomenclature for viruses with single-sense genomic ssDNA is similar to RNA nomenclature, in that the *template strand* for the viral mRNA is complementary to it (−), and the *coding strand* is a copy of it (+). However, several types of ssDNA and ssRNA viruses have genomes that are ambisense in that transcription can occur off both strands in a double-stranded replicative intermediate. Examples include geminiviruses, which are ssDNA plant viruses and arenaviruses, which are ssRNA viruses of animals.

Genome size varies greatly between species. The smallest viral genomes – the ssDNA circoviruses, family *Circoviridae* – code for only two proteins and have a genome size of only two kilobases; the largest–the pandoraviruses–have genome sizes of around two megabases which code for about 2500 proteins. Virus genes rarely have introns and often are arranged in the genome so that they overlap.

In general, RNA viruses have smaller genome sizes than DNA viruses because of a higher error-rate when replicating, and have a maximum upper size limit. Beyond this limit, errors in the genome when replicating render the virus useless or uncompetitive. To compensate for this, RNA viruses often have segmented genomes – the genome is split into smaller molecules – thus reducing the chance that an error in a single-component genome will incapacitate the entire genome. In contrast, DNA viruses generally have larger genomes because of the high fidelity of their replication enzymes. Single-strand DNA viruses are an exception to this rule, however, as mutation rates for these genomes can approach the extreme of the ssRNA virus case.

Genetic Mutation

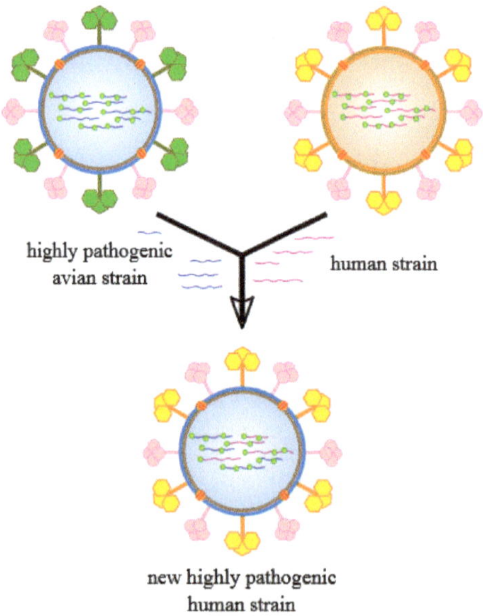

How antigenic shift, or reassortment, can result in novel and highly pathogenic strains of human flu

Viruses undergo genetic change by several mechanisms. These include a process called antigenic drift where individual bases in the DNA or RNA mutate to other bases. Most of these point mutations are "silent" – they do not change the protein that the gene encodes but others can confer evolutionary advantages such as resistance to antiviral drugs. Antigenic shift occurs when there is a major change in the genome of the virus. This can be a result of recombination or reassortment. When this happens with influenza viruses, pandemics might result. RNA viruses often exist as quasispecies or swarms of viruses of the same species but with slightly different genome nucleoside sequences. Such quasispecies are a prime target for natural selection.

Segmented genomes confer evolutionary advantages; different strains of a virus with a segmented genome can shuffle and combine genes and produce progeny viruses or (offspring) that have unique characteristics. This is called reassortment or *viral sex*.

Genetic recombination is the process by which a strand of DNA is broken and then joined to the end of a different DNA molecule. This can occur when viruses infect cells simultaneously and studies of viral evolution have shown that recombination has been rampant in the species studied. Recombination is common to both RNA and DNA viruses.

Replication Cycle

Viral populations do not grow through cell division, because they are acellular. Instead, they use the machinery and metabolism of a host cell to produce multiple copies of themselves, and they *assemble* in the cell.

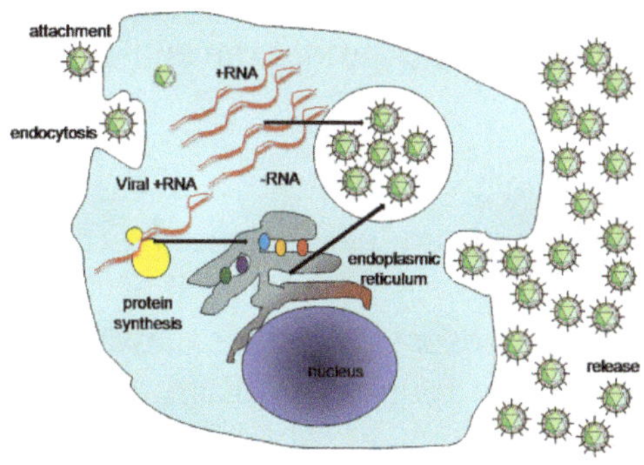

A typical virus replication cycle

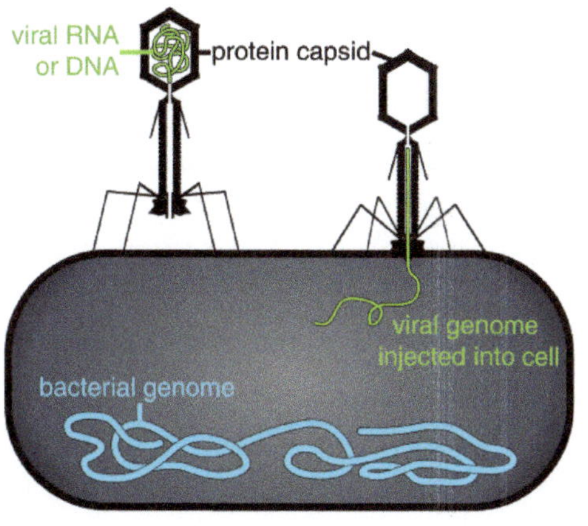

Some bacteriophages inject their genomes into bacterial cells (not to scale)

The life cycle of viruses differs greatly between species but there are six *basic* stages in the life cycle of viruses:

Attachment is a specific binding between viral capsid proteins and specific receptors on the host cellular surface. This specificity determines the host range of a virus. For example, HIV infects a limited range of human leucocytes. This is because its surface protein, gp120, specifically interacts with the CD4 molecule – a chemokine receptor – which is most commonly found on the surface of CD4+ T-Cells. This mechanism has evolved to favour those viruses that infect only cells in which they are capable of replication. Attachment to the receptor can induce the viral envelope protein to undergo changes that results in the fusion of viral and cellular membranes, or changes of non-enveloped virus surface proteins that allow the virus to enter.

Penetration follows attachment: Virions enter the host cell through receptor-mediated endocytosis or membrane fusion. This is often called viral entry. The infection of plant and fungal cells is different from that of animal cells. Plants have a rigid cell wall made of cellulose, and fungi one of chitin, so most viruses can get inside these cells only after trauma to the cell wall. However, nearly all plant viruses (such as tobacco mosaic virus) can also move directly from cell to cell, in the form of single-stranded nucleoprotein complexes, through pores called plasmodesmata. Bacteria, like plants, have strong cell walls that a virus must breach to infect the cell. However, given that bacterial cell walls are much less thick than plant cell walls due to their much smaller size, some viruses have evolved mechanisms that inject their genome into the bacterial cell across the cell wall, while the viral capsid remains outside.

Uncoating is a process in which the viral capsid is removed: This may be by degradation by viral enzymes or host enzymes or by simple dissociation; the end-result is the releasing of the viral genomic nucleic acid.

Replication of viruses involves primarily multiplication of the genome. Replication involves synthesis of viral messenger RNA (mRNA) from "early" genes (with exceptions for positive sense RNA viruses), viral protein synthesis, possible assembly of viral proteins, then viral genome replication mediated by early or regulatory protein expression. This may be followed, for complex viruses with larger genomes, by one or more further rounds of mRNA synthesis: "late" gene expression is, in general, of structural or virion proteins.

Assembly – Following the structure-mediated self-*assembly* of the virus particles, some modification of the proteins often occurs. In viruses such as HIV, this modification (sometimes called maturation) occurs *after* the virus has been released from the host cell.

Release – Viruses can be *released* from the host cell by lysis, a process that kills the cell by bursting its membrane and cell wall if present: This is a feature of many bacterial and some animal viruses. Some viruses undergo a lysogenic cycle where the viral genome is incorporated by genetic recombination into a specific place in the host's chromosome. The viral genome is then known as a "provirus" or, in the case of bacteriophages a "prophage". Whenever the host divides, the viral genome is also replicated. The viral genome is mostly silent within the host. However, at some point, the provirus or prophage may give rise to active virus, which may lyse the host cells. Enveloped viruses (e.g., HIV) typically are released from the host cell by budding. During this process the virus acquires its envelope, which is a modified piece of the host's plasma or other, internal membrane.

Genome Replication

The genetic material within virus particles, and the method by which the material is replicated, varies considerably between different types of viruses.

DNA Viruses

The genome replication of most DNA viruses takes place in the cell's nucleus. If the cell has the appropriate receptor on its surface, these viruses enter the cell sometimes by direct fusion with the cell membrane (e.g., herpesviruses) or – more usually – by receptor-mediated endocytosis. Most DNA viruses are entirely dependent on the host cell's DNA and RNA synthesising machinery, and RNA processing machinery. However, viruses with larger genomes may encode much of this machinery themselves. In eukaryotes the viral genome must cross the cell's nuclear membrane to access this machinery, while in bacteria it need only enter the cell.

RNA Viruses

Replication usually takes place in the cytoplasm. RNA viruses can be placed into four different groups depending on their modes of replication. The polarity (whether or not it can be used directly by ribosomes to make proteins) of single-stranded RNA viruses largely determines the replicative mechanism; the other major criterion is whether the genetic material is single-stranded or double-stranded. All RNA viruses use their own RNA replicase enzymes to create copies of their genomes.

Reverse Transcribing Viruses

These have ssRNA (*Retroviridae, Metaviridae, Pseudoviridae*) or dsDNA (*Caulimoviridae*, and *Hepadnaviridae*) in their particles. Reverse transcribing viruses with RNA genomes (retroviruses), use a DNA intermediate to replicate, whereas those with DNA genomes (pararetroviruses) use an RNA intermediate during genome replication. Both types use a reverse transcriptase, or RNA-dependent DNA polymerase enzyme, to carry out the nucleic acid conversion. Retroviruses integrate the DNA produced by reverse transcription into the host genome as a provirus as a part of the replication process; pararetroviruses do not, although integrated genome copies of especially plant pararetroviruses can give rise to infectious virus. They are susceptible to antiviral drugs that inhibit the reverse transcriptase enzyme, e.g. zidovudine and lamivudine. An example of the first type is HIV, which is a retrovirus. Examples of the second type are the *Hepadnaviridae*, which includes Hepatitis B virus.

Effects on the Host Cell

The range of structural and biochemical effects that viruses have on the host cell is extensive. These are called *cytopathic effects*. Most virus infections eventually result in the death of the host cell. The causes of death include cell lysis, alterations to the cell's surface membrane and apoptosis. Often cell death is caused by cessation of its normal activities because of suppression by virus-specific proteins, not all of which are components of the virus particle.

Some viruses cause no apparent changes to the infected cell. Cells in which the virus is latent and inactive show few signs of infection and often function normally. This causes persistent infections and the virus is often dormant for many months or years. This is often the case with herpes viruses. Some viruses, such as Epstein–Barr virus, can cause cells to proliferate without causing malignancy, while others, such as papillomaviruses, are established causes of cancer.

Host Range

Viruses are by far the most abundant biological entities on Earth and they outnumber all the others put together. They infect all types of cellular life including animals, plants, bacteria and fungi. However, different types of viruses can infect only a limited range of hosts and many are species-specific. Some, such as smallpox virus for example, can infect only one species – in this case humans, and are said to have a narrow host range. Other viruses, such as rabies virus, can infect different species of mammals and are said to have a broad range. The viruses that infect plants are harmless to animals, and most viruses that infect other animals are harmless to humans. The host range of some bacteriophages is limited to a single strain of bacteria and they can be used to trace the source of outbreaks of infections by a method called phage typing.

Classification

Classification seeks to describe the diversity of viruses by naming and grouping them on the basis of similarities. In 1962, André Lwoff, Robert Horne, and Paul Tournier were the first to develop a means of virus classification, based on the Linnaean hierarchical system. This system bases classification on phylum, class, order, family, genus, and species. Viruses were grouped according to their shared properties (not those of their hosts) and the type of nucleic acid forming their genomes. Later the International Committee on Taxonomy of Viruses was formed. However, viruses are not classified on the basis of phylum or class, as their small genome size and high rate of mutation makes it difficult to determine their ancestry beyond order. As such, the Baltimore classification is used to supplement the more traditional hierarchy.

ICTV Classification

The International Committee on Taxonomy of Viruses (ICTV) developed the current classification system and wrote guidelines that put a greater weight on certain virus properties to maintain family uniformity. A unified taxonomy (a universal system for classifying viruses) has been established. The 9th ICTV Report defines the concept of the virus species as the lowest taxon (group) in a branching hierarchy of viral taxa. However, at present only a small part of the total diversity of viruses has been studied, with analyses of samples from humans finding that about 20% of the virus sequences recovered have not been seen before, and samples from the environment, such as from seawater and ocean sediments, finding that the large majority of sequences are completely novel.

The general taxonomic structure is as follows:

- Order (*-virales*)
- Family (*-viridae*)
- Subfamily (*-virinae*)
- Genus (*-virus*)
- Species (*-virus*)

In the current (2013) ICTV taxonomy, 7 orders have been established, the *Caudovirales*, *Herpesvirales*, *Ligamenvirales*, *Mononegavirales*, *Nidovirales*, *Picornavirales*, and *Tymovirales*. The committee does not formally distinguish between subspecies, strains, and isolates. In total there are 7 orders, 103 families, 22 subfamilies, 455 genera, about 2,827 species and over 4,000 types yet unclassified.

Baltimore Classification

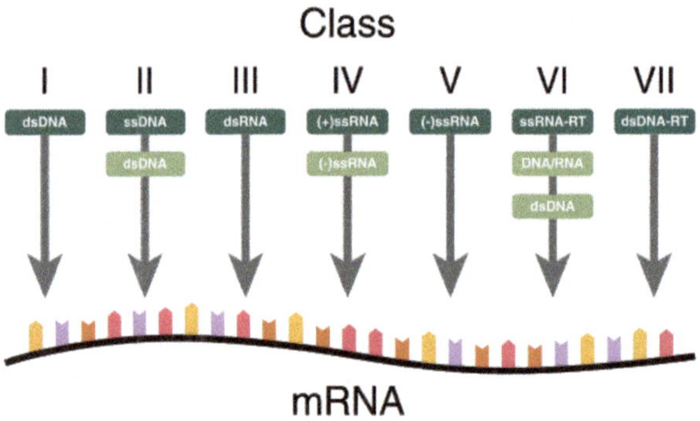

The Baltimore Classification of viruses is based on the method of viral mRNA synthesis

The Nobel Prize-winning biologist David Baltimore devised the Baltimore classification system. The ICTV classification system is used in conjunction with the Baltimore classification system in modern virus classification.

The Baltimore classification of viruses is based on the mechanism of mRNA production. Viruses must generate mRNAs from their genomes to produce proteins and replicate themselves, but different mechanisms are used to achieve this in each virus family. Viral genomes may be single-stranded (ss) or double-stranded (ds), RNA or DNA, and may or may not use reverse transcriptase (RT). In addition, ssRNA viruses may be either sense (+) or antisense (–). This classification places viruses into seven groups:

- I: dsDNA viruses (e.g. Adenoviruses, Herpesviruses, Poxviruses)
- II: ssDNA viruses (+ strand or "sense") DNA (e.g. Parvoviruses)
- III: dsRNA viruses (e.g. Reoviruses)
- IV: (+)ssRNA viruses (+ strand or sense) RNA (e.g. Picornaviruses, Togaviruses)
- V: (–)ssRNA viruses (– strand or antisense) RNA (e.g. Orthomyxoviruses, Rhabdoviruses)
- VI: ssRNA-RT viruses (+ strand or sense) RNA with DNA intermediate in life-cycle (e.g. Retroviruses)
- VII: dsDNA-RT viruses (e.g. Hepadnaviruses)

As an example of viral classification, the chicken pox virus, varicella zoster (VZV), belongs to the order *Herpesvirales*, family *Herpesviridae*, subfamily *Alphaherpesvirinae*, and genus *Varicellovirus*. VZV is in Group I of the Baltimore Classification because it is a dsDNA virus that does not use reverse transcriptase.

The complete set of viruses in an organism or habitat is called the virome; for example, all human viruses constitute the human virome.

Role in Human Disease

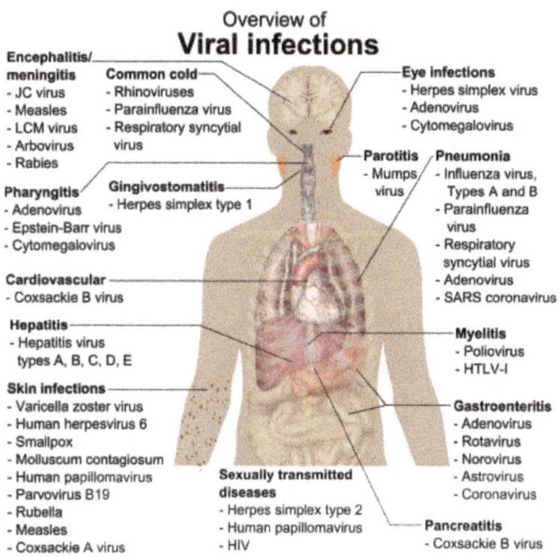

Overview of the main types of viral infection and the most notable species involved

Examples of common human diseases caused by viruses include the common cold, influenza, chickenpox, and cold sores. Many serious diseases such as Ebola virus disease, AIDS, avian influenza, and SARS are caused by viruses. The relative ability of viruses to cause disease is described in terms of virulence. Other diseases are under investigation to discover if they have a virus as the causative agent, such as the possible connection between human herpesvirus 6 (HHV6) and neurological diseases such as multiple sclerosis and chronic fatigue syndrome. There is controversy over whether the bornavirus, previously thought to cause neurological diseases in horses, could be responsible for psychiatric illnesses in humans.

Viruses have different mechanisms by which they produce disease in an organism, which depends largely on the viral species. Mechanisms at the cellular level primarily include cell lysis, the breaking open and subsequent death of the cell. In multicellular organisms, if enough cells die, the whole organism will start to suffer the effects. Although viruses cause disruption of healthy homeostasis, resulting in disease, they may exist relatively harmlessly within an organism. An example would include the ability of the herpes simplex virus, which causes cold sores, to remain in a dormant state within the human body. This is called latency and is a characteristic of the herpes viruses, including Epstein–Barr virus, which causes glandular fever, and varicella zoster virus, which causes chickenpox and shingles. Most people have been infected with at least one of these types of herpes virus. However, these latent viruses might sometimes be beneficial, as the presence of the virus can increase immunity against bacterial pathogens, such as *Yersinia pestis*.

Some viruses can cause lifelong or chronic infections, where the viruses continue to replicate in the body despite the host's defence mechanisms. This is common in hepatitis B virus and hepatitis C virus infections. People chronically infected are known as carriers, as they serve as reservoirs of infectious virus. In populations with a high proportion of carriers, the disease is said to be endemic.

Epidemiology

Viral epidemiology is the branch of medical science that deals with the transmission and control of virus infections in humans. Transmission of viruses can be vertical, which means from mother to child, or horizontal, which means from person to person. Examples of vertical transmission include hepatitis B virus and HIV, where the baby is born already infected with the virus. Another, more rare, example is the varicella zoster virus, which, although causing relatively mild infections in humans, can be fatal to the foetus and newborn baby.

Horizontal transmission is the most common mechanism of spread of viruses in populations. Transmission can occur when: body fluids are exchanged during sexual activity, e.g., HIV; blood is exchanged by contaminated transfusion or needle sharing, e.g., hepatitis C; exchange of saliva by mouth, e.g., Epstein–Barr virus; contaminated food or water is ingested, e.g., norovirus; aerosols containing virions are inhaled, e.g., influenza virus; and insect vectors such as mosquitoes penetrate the skin of a host, e.g., dengue. The rate or speed of transmission of viral infections depends on factors that include population density, the number of susceptible individuals, (i.e., those not immune), the quality of healthcare and the weather.

Epidemiology is used to break the chain of infection in populations during outbreaks of viral diseases. Control measures are used that are based on knowledge of how the virus is transmitted. It is important to find the source, or sources, of the outbreak and to identify the virus. Once the virus has been identified, the chain of transmission can sometimes be broken by vaccines. When vaccines are not available, sanitation and disinfection can be effective. Often, infected people are isolated from the rest of the community, and those that have been exposed to the virus are placed in quarantine. To control the outbreak of foot-and-mouth disease in cattle in Britain in 2001, thousands of cattle were slaughtered. Most viral infections of humans and other animals have incubation periods during which the infection causes no signs or symptoms. Incubation periods for viral diseases range from a few days to weeks, but are known for most infections. Somewhat overlapping, but mainly following the incubation period, there is a period of communicability — a time when an infected individual or animal is contagious and can infect another person or animal. This, too, is known for many viral infections, and knowledge of the length of both periods is important in the control of outbreaks. When outbreaks cause an unusually high proportion of cases in a population, community, or region, they are called epidemics. If outbreaks spread worldwide, they are called pandemics.

Epidemics and Pandemics

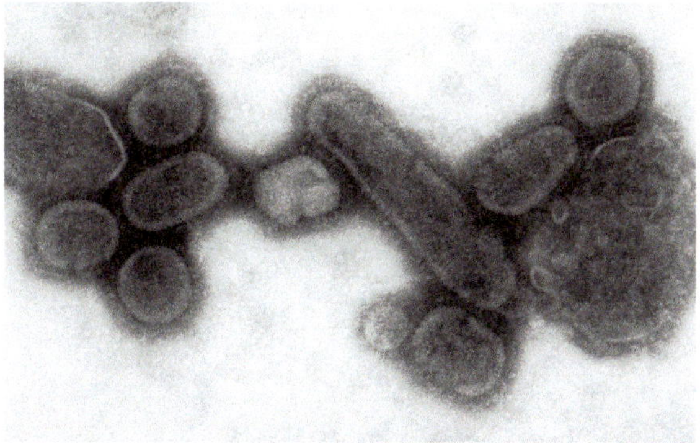

Transmission electron microscope image of a recreated 1918 influenza virus

Native American populations were devastated by contagious diseases, in particular, smallpox, brought to the Americas by European colonists. It is unclear how many Native Americans were killed by foreign diseases after the arrival of Columbus in the Americas, but the numbers have been estimated to be close to 70% of the indigenous population. The damage done by this disease significantly aided European attempts to displace and conquer the native population.

A pandemic is a worldwide epidemic. The 1918 flu pandemic, which lasted until 1919, was a category 5 influenza pandemic caused by an unusually severe and deadly influenza A virus. The victims were often healthy young adults, in contrast to most influenza outbreaks, which predominantly affect juvenile, elderly, or otherwise-weakened patients. Older estimates say it killed 40–50 million people, while more recent research suggests that it may have killed as many as 100 million people, or 5% of the world's population in 1918.

Most researchers believe that HIV originated in sub-Saharan Africa during the 20th century; it is now a pandemic, with an estimated 38.6 million people now living with the disease worldwide. The Joint United Nations Programme on HIV/AIDS (UNAIDS) and the World Health Organization (WHO) estimate that AIDS has killed more than 25 million people since it was first recognised on 5 June 1981, making it one of the most destructive epidemics in recorded history. In 2007 there were 2.7 million new HIV infections and 2 million HIV-related deaths.

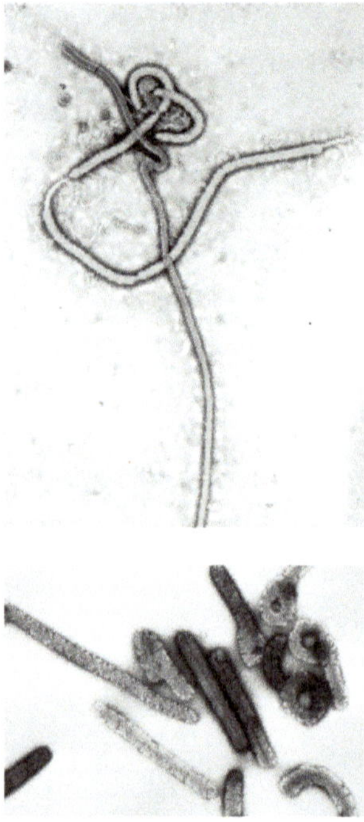

Ebola (top) and Marburg viruses (bottom)

Several highly lethal viral pathogens are members of the *Filoviridae*. Filoviruses are filament-like viruses that cause viral hemorrhagic fever, and include ebolaviruses and marburgviruses. Marburg virus, first discovered in 1967, attracted widespread press attention in April 2005 for an outbreak in Angola. Ebola Virus Disease has also caused intermittent outbreaks with high mortality rates since 1976 when it was first identified. The worst and most recent one is the West Africa epidemic.

Cancer

Viruses are an established cause of cancer in humans and other species. Viral cancers occur only in a minority of infected persons (or animals). Cancer viruses come from a range of virus families, including both RNA and DNA viruses, and so there is no single type of "oncovirus" (an obsolete term originally used for acutely transforming retroviruses). The development of cancer is determined by a variety of factors such as host immunity and mutations in the host. Viruses accepted to cause human cancers include some genotypes of human papillomavirus, hepatitis B virus, hepatitis C virus, Epstein–Barr virus, Kaposi's sarcoma-associated herpesvirus and human T-lymphotropic virus. The most recently discovered human cancer virus is a polyomavirus (Merkel cell polyomavirus) that causes most cases of a rare form of skin cancer called Merkel cell carcinoma. Hepatitis viruses can develop into a chronic viral infection that leads to liver cancer. Infection by human T-lymphotropic virus can lead to tropical spastic paraparesis and adult T-cell leukaemia. Human papillomaviruses are an established cause of cancers of cervix, skin, anus, and penis. Within the *Herpesviridae*, Kaposi's sarcoma-associated herpesvirus causes Kaposi's sarcoma and body-cavity lymphoma, and Epstein–Barr virus causes Burkitt's lymphoma, Hodgkin's lymphoma, B lymphoproliferative disorder, and nasopharyngeal carcinoma. Merkel cell polyomavirus closely related to SV40 and mouse polyomaviruses that have been used as animal models for cancer viruses for over 50 years.

Host Defence Mechanisms

The body's first line of defence against viruses is the innate immune system. This comprises cells and other mechanisms that defend the host from infection in a non-specific manner. This means that the cells of the innate system recognise, and respond to, pathogens in a generic way, but, unlike the adaptive immune system, it does not confer long-lasting or protective immunity to the host.

RNA interference is an important innate defence against viruses. Many viruses have a replication strategy that involves double-stranded RNA (dsRNA). When such a virus infects a cell, it releases its RNA molecule or molecules, which immediately bind to a protein complex called a dicer that cuts the RNA into smaller pieces. A biochemical pathway – the RISC complex, is activated, which ensures cell survival by degrading the viral mRNA. Rotaviruses have evolved to avoid this defence mechanism by not uncoating fully inside the cell, and releasing newly produced mRNA through pores in the particle's inner capsid. Their genomic dsRNA remains protected inside the core of the virion.

When the adaptive immune system of a vertebrate encounters a virus, it produces specific antibodies that bind to the virus and often render it non-infectious. This is called humoral immunity. Two types of antibodies are important. The first, called IgM, is highly effective at neutralising viruses but is produced by the cells of the immune system only for a few weeks. The second, called IgG, is produced indefinitely. The presence of IgM in the blood of the host is used to test for acute infection, whereas IgG indicates an infection sometime in the past. IgG antibody is measured when tests for immunity are carried out.

Antibodies can continue to be an effective defence mechanism even after viruses have managed to gain entry to the host cell. A protein that is in cells, called TRIM21, can attach to the antibodies on the surface of the virus particle. This primes the subsequent destruction of the virus by the enzymes of the cell's proteosome system.

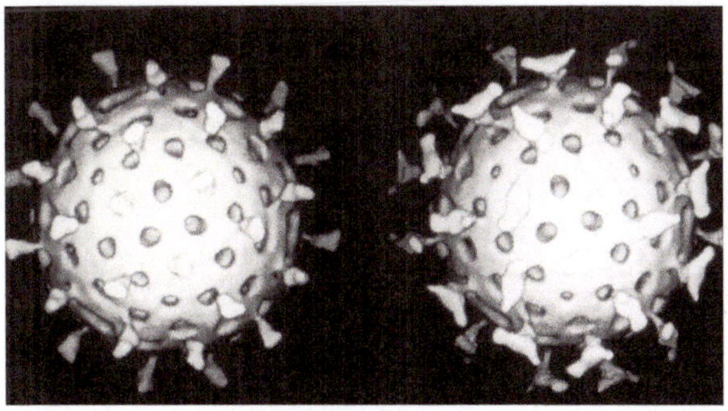

Two rotaviruses: the one on the right is coated with antibodies that prevent its attachment to cells and infecting them.

A second defence of vertebrates against viruses is called cell-mediated immunity and involves immune cells known as T cells. The body's cells constantly display short fragments of their proteins on the cell's surface, and, if a T cell recognises a suspicious viral fragment there, the host cell is destroyed by *killer T* cells and the virus-specific T-cells proliferate. Cells such as the macrophage are specialists at this antigen presentation. The production of interferon is an important host defence mechanism. This is a hormone produced by the body when viruses are present. Its role in immunity is complex; it eventually stops the viruses from reproducing by killing the infected cell and its close neighbours.

Not all virus infections produce a protective immune response in this way. HIV evades the immune system by constantly changing the amino acid sequence of the proteins on the surface of the virion. This is known as "escape mutation" as the viral epitopes escape recognition by the host immune response. These persistent viruses evade immune control by sequestration, blockade of antigen presentation, cytokine resistance, evasion of natural killer cell activities, escape from apoptosis, and antigenic shift. Other viruses, called *neurotropic viruses*, are disseminated by neural spread where the immune system may be unable to reach them.

Prevention and Treatment

Because viruses use vital metabolic pathways within host cells to replicate, they are difficult to eliminate without using drugs that cause toxic effects to host cells in general. The most effective medical approaches to viral diseases are vaccinations to provide immunity to infection, and antiviral drugs that selectively interfere with viral replication.

Vaccines

Vaccination is a cheap and effective way of preventing infections by viruses. Vaccines were used to prevent viral infections long before the discovery of the actual viruses. Their use has resulted in a dramatic decline in morbidity (illness) and mortality (death) associated with viral infections such as polio, measles, mumps and rubella. Smallpox infections have been eradicated. Vaccines are available to prevent over thirteen viral infections of humans, and more are used to prevent viral infections of animals. Vaccines can consist of live-attenuated or killed viruses, or viral proteins (antigens). Live vaccines contain weakened forms of the virus, which do not cause the disease but, nonetheless, confer immunity. Such viruses are called attenuated. Live vaccines can be dangerous when given to people with a weak immunity (who are described as immunocompromised), because in these people, the weakened virus can cause the original disease. Biotechnology and genetic engineering techniques are used to produce subunit vaccines. These vaccines use only the capsid proteins of the virus. Hepatitis B vaccine is an example of this type of vaccine. Subunit vaccines are safe for immunocompromised patients because they cannot cause the disease. The yellow fever virus vaccine, a live-attenuated strain called 17D, is probably the safest and most effective vaccine ever generated.

Antiviral Drugs

Antiviral drugs are often nucleoside analogues (fake DNA building-blocks), which viruses mistakenly incorporate into their genomes during replication. The life-cycle of the virus is then halted because the newly synthesised DNA is inactive. This is because these analogues lack the hydroxyl groups, which, along with phosphorus atoms, link together to form the strong "backbone" of the DNA molecule. This is called DNA chain termination. Examples of nucleoside analogues are aciclovir for Herpes simplex virus infections and lamivudine for HIV and Hepatitis B virus infections. Aciclovir is one of the oldest and most frequently prescribed antiviral drugs. Other antiviral drugs in use target different stages of the viral life cycle. HIV is dependent on a proteolytic enzyme called the HIV-1 protease for it to become fully infectious. There is a large class of drugs called protease inhibitors that inactivate this enzyme.

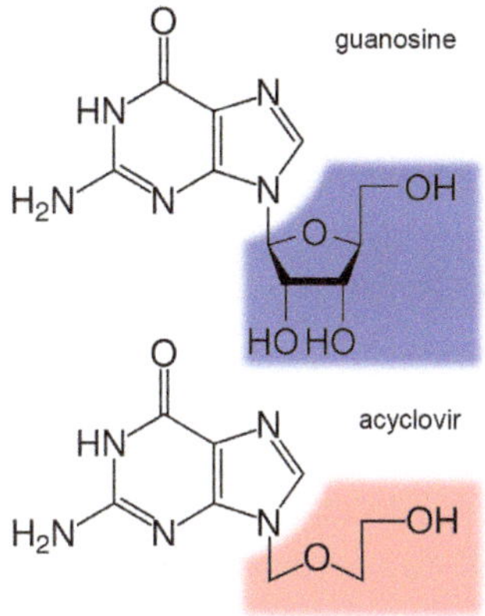

The structure of the DNA base guanosine and the antiviral drug acyclovir

Hepatitis C is caused by an RNA virus. In 80% of people infected, the disease is chronic, and without treatment, they are infected for the remainder of their lives. However, there is now an effective treatment that uses the nucleoside analogue drug ribavirin combined with interferon. The treatment of chronic carriers of the hepatitis B virus by using a similar strategy using lamivudine has been developed.

Infection in other Species

Viruses infect all cellular life and, although viruses occur universally, each cellular species has its own specific range that often infect only that species. Some viruses, called satellites, can replicate only within cells that have already been infected by another virus.

Animal Viruses

Viruses are important pathogens of livestock. Diseases such as foot-and-mouth disease and bluetongue are caused by viruses. Companion animals such as cats, dogs, and horses, if not vaccinated, are susceptible to serious viral infections. Canine parvovirus is caused by a small DNA virus and infections are often fatal in pups. Like all invertebrates, the honey bee is susceptible to many viral infections. However, most viruses co-exist harmlessly in their host and cause no signs or symptoms of disease.

Plant Viruses

Peppers infected by mild mottle virus

There are many types of plant virus, but often they cause only a loss of yield, and it is not economically viable to try to control them. Plant viruses are often spread from plant to plant by organisms, known as *vectors*. These are normally insects, but some fungi, nematode worms, and single-celled organisms have been shown to be vectors. When control of plant virus infections is considered economical, for perennial fruits, for example, efforts are concentrated on killing the vectors and removing alternate hosts such as weeds. Plant viruses cannot infect humans and other animals because they can reproduce only in living plant cells.

Plants have elaborate and effective defence mechanisms against viruses. One of the most effective is the presence of so-called resistance (R) genes. Each R gene confers resistance to a particular virus by triggering localised areas of cell death around the infected cell, which can often be seen with the unaided eye as large spots. This stops the infection from spreading. RNA interference is also an effective defence in plants. When they are infected, plants often produce natural disinfectants that kill viruses, such as salicylic acid, nitric oxide, and reactive oxygen molecules.

Plant virus particles or virus-like particles (VLPs) have applications in both biotechnology and nanotechnology. The capsids of most plant viruses are simple and robust structures and can be produced in large quantities either by the infection of plants or by expression in a variety of heterologous systems. Plant virus particles can be modified genetically and chemically to encapsulate foreign material and can be incorporated into supramolecular structures for use in biotechnology.

Bacterial Viruses

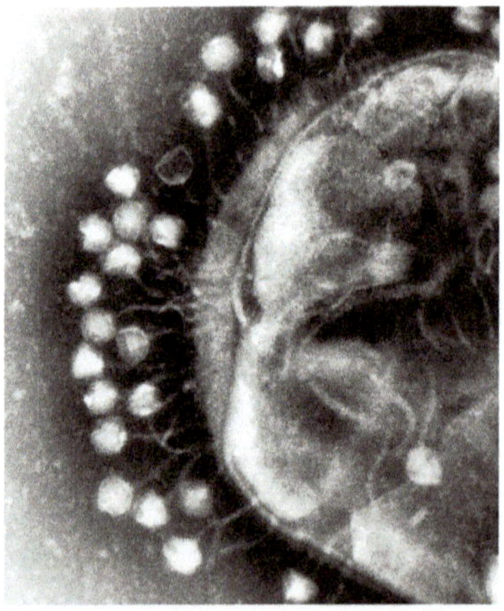

Transmission electron micrograph of multiple bacteriophages attached to a bacterial cell wall

Bacteriophages are a common and diverse group of viruses and are the most abundant form of biological entity in aquatic environments – there are up to ten times more of these viruses in the oceans than there are bacteria, reaching levels of 250,000,000 bacteriophages per millilitre of seawater. These viruses infect specific bacteria by binding to surface receptor molecules and then entering the cell. Within a short amount of time, in some cases just minutes, bacterial polymerase starts translating viral mRNA into protein. These proteins go on to become either new virions within the cell, helper proteins, which help assembly of new virions, or proteins involved in cell lysis. Viral enzymes aid in the breakdown of the cell membrane, and, in the case of the T4 phage, in just over twenty minutes after injection over three hundred phages could be released.

The major way bacteria defend themselves from bacteriophages is by producing enzymes that destroy foreign DNA. These enzymes, called restriction endonucleases, cut up the viral DNA that bacteriophages inject into bacterial cells. Bacteria also contain a system that uses CRISPR sequences to retain fragments of the genomes of viruses that the bacteria have come into contact with in the past, which allows them to block the virus's replication through a form of RNA interference. This genetic system provides bacteria with acquired immunity to infection.

Archaean Viruses

Some viruses replicate within archaea: these are double-stranded DNA viruses with unusual and sometimes unique shapes. These viruses have been studied in most detail in the thermophilic archaea, particularly the orders Sulfolobales and Thermoproteales. Defences against these viruses involve RNA interference from repetitive DNA sequences within archaean genomes that are related to the genes of the viruses. Most archaea have CRISPR–Cas systems as an adaptive defence against viruses. These enable archaea to retain sections of viral DNA, which are then used to target and eliminate subsequent infections by the virus using a process similar to RNA interference.

Role in Aquatic Ecosystems

A teaspoon of seawater contains about one million viruses. Most of these are bacteriophages, which are harmless to plants and animals, and are in fact essential to the regulation of saltwater and freshwater ecosystems. They infect and destroy bacteria in aquatic microbial communities, and are the most important mechanism of recycling carbon in the marine environment. The organic molecules released from the dead bacterial cells stimulate fresh bacterial and algal growth. Viral activity may also contribute to the biological pump, the process whereby carbon is sequestered in the deep ocean.

Microorganisms constitute more than 90% of the biomass in the sea. It is estimated that viruses kill approximately 20% of this biomass each day and that there are 15 times as many viruses in the oceans as there are bacteria and archaea. Viruses are the main agents responsible for the rapid destruction of harmful algal blooms, which often kill other marine life. The number of viruses in the oceans decreases further offshore and deeper into the water, where there are fewer host organisms.

Like any organism, marine mammals are susceptible to viral infections. In 1988 and 2002, thousands of harbour seals were killed in Europe by phocine distemper virus. Many other viruses, including caliciviruses, herpesviruses, adenoviruses and parvoviruses, circulate in marine mammal populations.

Role in Evolution

Viruses are an important natural means of transferring genes between different species, which increases genetic diversity and drives evolution. It is thought that viruses played a central role in the early evolution, before the diversification of bacteria, archaea and eukaryotes, at the time of the last universal common ancestor of life on Earth. Viruses are still one of the largest reservoirs of unexplored genetic diversity on Earth.

Applications

Life Sciences and Medicine

Viruses are important to the study of molecular and cell biology as they provide simple systems that can be used to manipulate and investigate the functions of cells. The study and use of viruses have provided valuable information about aspects of cell biology. For example, viruses have been useful in the study of genetics and helped our understanding of the basic mechanisms of molecular genetics, such as DNA replication, transcription, RNA processing, translation, protein transport, and immunology.

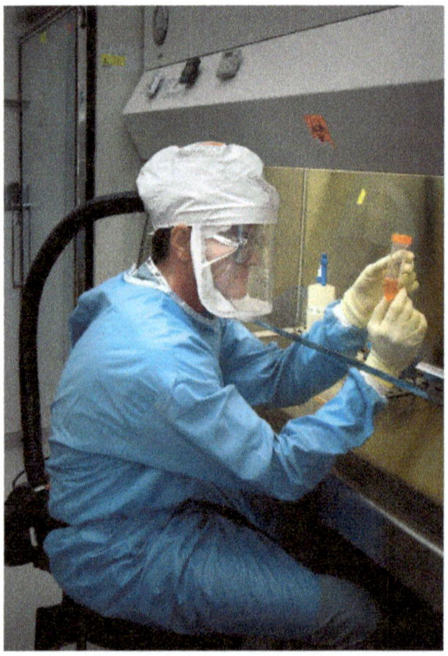

Scientist studying the H5N1 influenza virus

Geneticists often use viruses as vectors to introduce genes into cells that they are studying. This is useful for making the cell produce a foreign substance, or to study the effect of introducing a new gene into the genome. In similar fashion, virotherapy uses viruses as vectors to treat various diseases, as they can specifically target cells and DNA. It shows promising use in the treatment of cancer and in gene therapy. Eastern European scientists have used phage therapy as an alternative to antibiotics for some time, and interest in this approach is increasing, because of the high level of antibiotic resistance now found in some pathogenic bacteria. Expression of heterologous proteins by viruses is the basis of several manufacturing processes that are currently being used for the production of various proteins such as vaccine antigens and antibodies. Industrial processes have been recently developed using viral vectors and a number of pharmaceutical proteins are currently in pre-clinical and clinical trials.

Virotherapy

Virotherapy involves the use of genetically modified viruses to treat diseases. Viruses have been modified by scientists to reproduce in cancer cells and destroy them but not infect healthy cells. Talimogene laherparepvec (T-VEC), for example, is a modified herpes simplex virus that has had a gene, which is required for viruses to replicate in healthy cells, deleted and replaced with a human gene (GM-CSF) that stimulates immunity. When this virus infects cancer cells, it destroys them and in doing so the presence the GM-CSF gene attracts dendritic cells from the surrounding tissues of the body. The dendritic cells process the dead cancer cells and present components of them to other cells of the immune system. Having completed successful clinical trials, this virus is expected to gain approval for the treatment of a skin cancer called melanoma in late 2015. Viruses that have been reprogrammed to kill cancer cells are called oncolytic viruses.

Materials Science and Nanotechnology

Current trends in nanotechnology promise to make much more versatile use of viruses. From the viewpoint of a materials scientist, viruses can be regarded as organic nanoparticles. Their surface carries specific tools designed to cross the barriers of their host cells. The size and shape of viruses, and the number and nature of the functional groups on their surface, is precisely defined. As such, viruses are commonly used in materials science as scaffolds for covalently linked surface modifications. A particular quality of viruses is that they can be tailored by directed evolution. The powerful techniques developed by life sciences are becoming the basis of engineering approaches towards nanomaterials, opening a wide range of applications far beyond biology and medicine.

Because of their size, shape, and well-defined chemical structures, viruses have been used as templates for organising materials on the nanoscale. Recent examples include work at the Naval Research Laboratory in Washington, D.C., using Cowpea mosaic virus (CPMV) particles to amplify signals in DNA microarray based sensors. In this application, the virus particles separate the fluorescent dyes used for signalling to prevent the formation of non-fluorescent dimers that act as quenchers. Another example is the use of CPMV as a nanoscale breadboard for molecular electronics.

Synthetic Viruses

Many viruses can be synthesised de novo ("from scratch") and the first synthetic virus was created in 2002. Although somewhat of a misconception, it is not the actual virus that is synthesised, but rather its DNA genome (in case of a DNA virus), or a cDNA copy of its genome (in case of RNA viruses). For many virus families the naked synthetic DNA or RNA (once enzymatically converted back from the synthetic cDNA) is infectious when introduced into a cell. That is, they contain all the necessary information to produce new viruses. This technology is now being used to investigate novel vaccine strategies. The ability to synthesise viruses has far-reaching consequences, since viruses can no longer be regarded as extinct, as long as the information of their genome sequence is known and permissive cells are available. As of March 2014, the full-length genome sequences of 3843 different viruses, including smallpox, are publicly available in an online database maintained by the National Institutes of Health.

Weapons

The ability of viruses to cause devastating epidemics in human societies has led to the concern that viruses could be weaponised for biological warfare. Further concern was raised by the successful recreation of the infamous 1918 influenza virus in a laboratory. Smallpox virus devastated numerous societies throughout history before its eradication. There are only two centres in the world that are authorised by the WHO to keep stocks of smallpox virus: the State Research Center of Virology and Biotechnology VECTOR in Russia and the Centers for Disease Control and Prevention in the United States. Fears that it may be used as a weapon may not be totally unfounded. As the vaccine for smallpox sometimes had severe side-effects, it is no longer used routinely in any country. Thus, much of the modern human population has almost no established resistance to smallpox, and would be vulnerable to the virus.

Virus Diversity

Viruses are minute, non-living entities that copy themselves once inside the living host cells. All living organisms (animals, plants, fungi, and bacteria) have viruses that infect them. Typically viruses are made up of coat (or capsid) that protects its information molecule (RNA or DNA); these information molecules contain the blue prints for making more virus. The viruses are highly diverse in their shape, size, genetic information, and infectivity. Viruses are all around us, on an average a human body encounters billion virus particles every day. Our intestinal, respiratory, and urogenital tract are reservoirs for many different kinds of viruses, it is astonishing that with such constant exposure, there is little or no impact of these organisms in human health. The host defense mechanism is quite strong to remove all these in normal condition, while they cause many nasty diseases only when the person is immune-compromised . Although viruses have a limited host range but sometimes they may jump the species barrier and causes fatal disease, recent spread of swine influenza is an ideal example of such kind of spread.

The epidemic viruses, such as influenza and severe acute respiratory syndrome (SARS), cause diseases that rapidly spread to a large human population within no time, and seem to attract more scientific and public attention than do endemic viruses, which are continually present in a particular population.

Virology as a discipline is merely 100 years old and the way it expanded in this small period of time is rampant. To group the new emerging viruses in a specific group by specifying certain parameters was initiated in 1966 when international committee on the taxonomy of viruses (ICTV) was formed with the aim to classify the viruses. The ICTV has adopted a norm for the description of the viruses. Name for genera, subfamilies, families, and orders must all be a single word, ending with the suffixes -virus , -virinae , -viridae , and -virales respectively. In written usage, the name should be capitalized and italicized.

Viruses are obligate parasite which means their absolute dependence on living host system. This property of virus made it a valuable tool to study cell functions and its biology. Adenovirus is an example of DNA virus that enters the host nucleus but remains separated from the host genome and at the same time use host cell machinery for its replication. On the other hand influenza is a RNA virus that carries its own enzyme to replicate its genome while the viral proteins are synthesized by using the host cell machinery. Human immunodeficiency virus (HIV) is a retrovirus ; it contains RNA as a genetic material but it converts into DNA after entering the host cell by an enzyme called reverse transcriptase . It also contains enzymes in its virion namely, integrase and viral protease which helps HIV during maturation process inside the infected cells. Outer surface of HIV virion contains two surface glycoproteins called as gp120 and gp41 which helps in the attachment of virus to the cell surface.

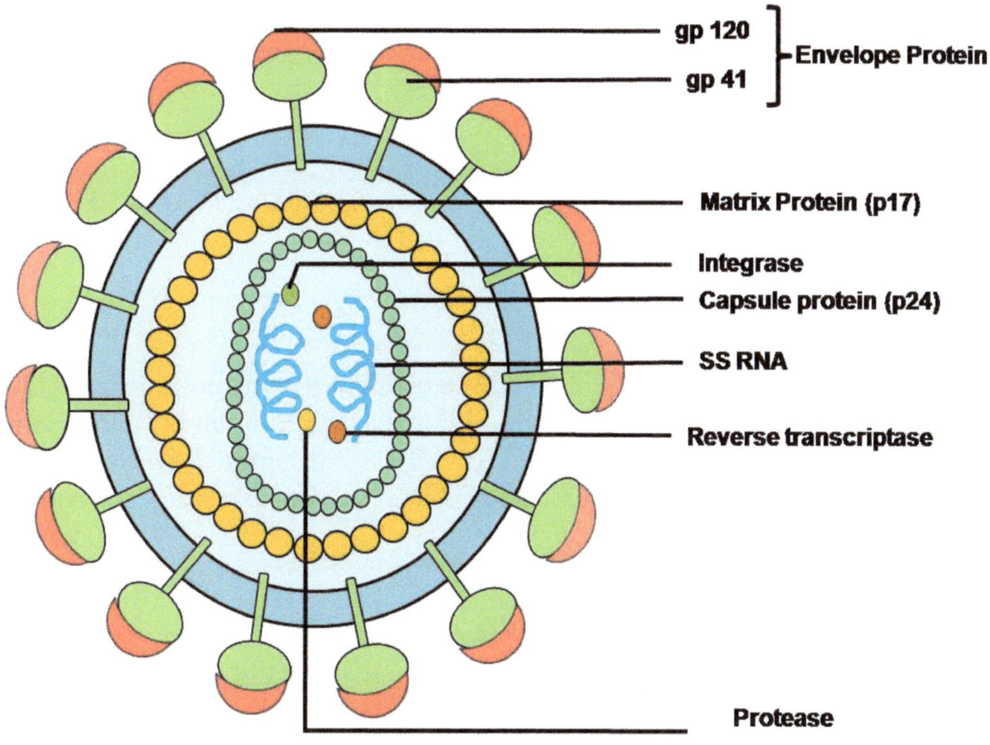

Schematic diagram of HIV

Obligate Parasite

An obligate parasite or holoparasite, is a parasitic organism that cannot complete its life-cycle without exploiting a suitable host. If an obligate parasite cannot obtain a host it will fail to reproduce. This is opposed to a facultative parasite, which can act as a parasite but does not rely on its host to continue its life-cycle. Obligate parasites have evolved a variety of parasitic strategies to exploit their hosts.

It is advantageous for the parasite to preserve the health of their host when this is compatible with their nutritional and reproductive requirements, except when the death of the host is necessary for transmission.

Species

Obligate parasitism is exhibited in a range of organisms, with examples in viruses, bacteria, fungi, plants, and animals. They are unable to complete their development without passing through at least one parasitic stage which is necessary to their life-cycle.

Whether one regards viruses as living organisms or not, they cannot reproduce except by means of resources within living cells, and accordingly, it is convenient and customary to regard them as obligate intracellular parasites.

Among the Vespidae family, *Vespula austriaca* is an example of an obligate parasite; its common host is *Vespula acadica*. In the genus *Bombus*, *B. bohemicus* is an obligate parasite of *B. locurum*, *B. cryptarum*, and *B. terrestris*.

Host-parasite Interaction

Life-cycle

Parasitic life-cycles involve the exploitation of at least one host. Parasites that infect a single species are said to have direct life-cycles. For example, the hookworm species *Necator americanus*. Parasites that infect more than one host are said to have a complex or indirect life-cycle. For example, the Malaria plasmodium.

Intermediate or Final Host

An intermediate or secondary host is exploited by the parasite only for a short transition period. A final or primary host is exploited by the parasite and is the only location in which the parasite is able to reach maturity and if possible, reproduce sexually. For example, *Ribeiroia ondatrae* uses Ramshorn snail as its first intermediate host, amphibians and fish as second intermediate hosts and birds as definitive hosts.

Parasitic Permanence

Obligate parasites may not necessarily spend all of their time behaving as parasites. When a parasite is permanent, a number of generations occur in or on the host of an infested individual. Head lice are an example of this. Temporary parasites are organisms whose parasitic mode of life is limited to a few or even one stage of development. An example of this is the larval stage of harvest mites, while the adult stage is non-parasitic.

Location on Host

The parasite may live outside of the host ectoparasite for example, a tick. Alternatively, the parasite may live within the host endoparasite for example, the fluke. A special alternative whereby the obligate parasite does not live directly in or on the host, but rather acts at a distance. For example, a cuckoo which hatches and is raised by non-relatives, is known as a brood parasite.

Invasion Strategies

In order to establish infestation in a susceptible host, obligate parasites must evade defences before, during and after entry into the host. Due to the wide range of obligate parasite types, it is impossible to identify a general invasion strategy. Intracellular parasites use various strategies to invade cells and subvert cellular signalling pathways. Most bacteria and viruses undergo passive uptake, where they rely on the host cell for uptake. However, apicomplexans engage in active entry. One obligate wasp parasite, *Polistes atrimandibularis*, infiltrates their hosts' colony by modifying their chemical signature to match that of the hosts'. This tricks the host wasps into thinking the parasite is one of their own.

Evasion of Host Defences

A number of obligate intracellular parasites have evolved mechanisms for evading their hosts cellular defences, including the ability to survive in distinct cellular compartments. One of the mechanisms that hosts employ in their attempt to reduce the replication and spread of pathogens is apoptosis (programmed cell death). Some obligate parasites have developed ways to suppress this phenomena, for example *Toxoplasma gondii* although the mechanism is not yet fully understood.

Manipulation of Host Behaviour

Changes in a host's behaviour following infection with obligate parasites are extremely common. Unusual behaviour observed in infected individuals is noted, and if its complexity suggests that this behaviour will benefit the transmission of the parasite, then this is said to be an example of adaptive manipulation. However, there is a difficulty in demonstrating changes in behaviour are the result of a selective process favouring transmission of the parasite. It has been suggested that these changes may merely be a side-effect of infection. Most behaviour changes have not been demonstrated to lead to fitness gains in either the host or the parasite. An example of this behaviour is the attraction of rats to cat urine after infection with *Toxoplasma gondii*.

Extended Phenotype

In some cases the behaviour we observe in an organism is not due to the expression of their genes, but rather to the genes of parasites infecting them. This behaviour is an extended phenotype.

Evolution of Host Behaviour Manipulation

Three main evolutionary routes have been suggested for the appearance of host behaviour manipulation by parasites. The first is a *parasite driven* scenario Of manipulation, while the second and third are *host driven* scenarios of manipulation.

1. Manipulation *sensu stricto* (extended phenotype- abhorrent behaviour displayed by parasitised hosts results from the expression of the parasites genes) this capacity could have been the product of natural selection in an ancestral parasite with the trait.
2. The mafia-like strategy- retaliation for non-compliance (eg.great spotted cuckoo and magpie) magpies that eject the cuckoos eggs from their nest suffer a much greater rate of cuckoo predation.
3. The exploitation of compensatory responses induce host compensatory responses since these may at least partially match with the transmission routes of parasites. E.g. the sexually transmitted ectoparasite *Chrysomelobia labidomerae*, parasitizing the leaf beetle host *Labidomera clivicollis~* infected males exhibit increased sexual behaviour and as a result enhance inter- and intra- sexual contacts (copulation and competition) which provide more opportunities for parasite transmission.

It has been suggested that extended phenotype behaviours are not adaptive, but are Exaptative. While they may have a benefit for the parasitic organism, they did not arise with the intention of this benefit.

Parasitic Mimicry in Brood Parasites

The cowbird and cuckoo require the nests and parental care of other passerines in order for their young to fledge. These are known as brood parasites. The parasitic bird species mimics egg patterns and colours of the host species, which reduces egg rejection. The chicks of some species are able to manipulate host behaviour by making rapid calls that mimic the sound made by up to four of the host chicks. Mimicry of the host species also occurs in the paper wasp species *Polistes semenowi* and *Polistes sulcifer* and the bumblebee species *Bombus bohemicus*, with the parasite changing its proportions of cuticular hydrocarbons, species- and colony-specific identifying chemicals, to match that of the usurped host species.

It is proposed that this mimicry has evolved through two processes: either as coevolutionary responses to host defences against brood parasites or modifying pre-existing host provisioning strategies. Competition between the parasite and host young for parental resources might lead to exaggeration of the aspects of the signal that most effectively exploit host parents. The parasitic young are likely to experience stronger selection for exaggerated signals than host young, because they are unrelated to the other chicks in the nest and therefore under selection to behave more selfishly.

Evolution of Obligate Parasitism

Current theory in evolutionary biology indicates that host-parasite relationships may evolve towards equilibrial states of severe disease. This differs from the conventional belief that commensalism is the ideal equilibrium for both the host and parasite.

Tobacco Mosaic Virus

Tobacco mosaic virus (TMV) is a positive-sense single stranded RNA virus that infects a wide range of plants, especially tobacco and other members of the family Solanaceae. The infection causes characteristic patterns, such as "mosaic"-like mottling and discoloration on the leaves (hence the name). TMV was the first virus ever to be discovered. Although it was known from the late 19th century that an infectious disease was damaging tobacco crops, it was not until 1930 that the infectious agent was determined to be a virus.

History

In 1886, Adolf Mayer first described the tobacco mosaic disease that could be transferred between plants, similar to bacterial infections. In 1892, Dmitri Ivanovsky gave the first concrete evidence for the existence of a non-bacterial infectious agent, showing that infected sap remained infectious even after filtering through finest Chamberland filters. Later, in 1903, Ivanovsky published a paper describing abnormal crystal intracellular inclusions in the host cells of the affected tobacco plants and argued the connection between these inclusions and the infectious agent. However, Ivanovsky remained rather convinced, despite repeated failures to produce evidence, that the causal agent was an unculturable bacterium, too small to be retained on the employed Chamberland filters and to be detected in the light microscope. In 1898, Martinus Beijerinck independently replicated Ivanovsky's filtration experiments and then showed that the infectious agent was able to reproduce and multiply in the host cells of the tobacco plant. Beijerinck coined the term of "virus" to indicate that the causal agent of tobacco mosaic disease was of non-bacterial nature. Tobacco mosaic virus was the first virus to be crystallized. It was achieved by Wendell Meredith Stanley in 1935 who also showed that TMV remains active even after crystallization. For his work, he was awarded 1/3 of the Nobel Prize in Chemistry in 1946, even though it was later shown some of his conclusions (in particular, that the crystals were pure protein, and assembled by autocatalysis) were incorrect. The first electron microscopical images of TMV were made in 1939 by Gustav Kausche, Edgar Pfankuch and Helmut Ruska – the brother of Nobel Prize winner Ernst Ruska. In 1955, Heinz Fraenkel-Conrat and Robley Williams showed that purified TMV RNA and its capsid (coat) protein assemble by themselves to functional viruses, indicating that this is the most stable structure (the one with the lowest free energy). The crystallographer Rosalind Franklin worked for Stanley for about a month at Berkeley, and later designed and built a model of TMV for the 1958 World's Fair at Brussels. In 1958, she speculated that the virus was hollow, not solid, and hypothesized that the RNA of TMV is single-stranded. This conjecture was proven to be correct after her death and is now known to be the + strand. The investigations of tobacco mosaic disease and subsequent discovery of its viral nature were instrumental in the establishment of the general concepts of virology.

Structure

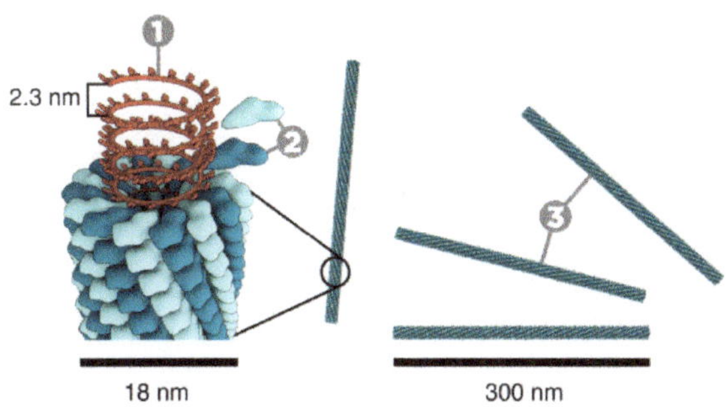

Schematic model of TMV: 1. nucleic acid (RNA), 2. capsomer protein (protomer), 3. capsid

Tobacco mosaic virus coat protein	
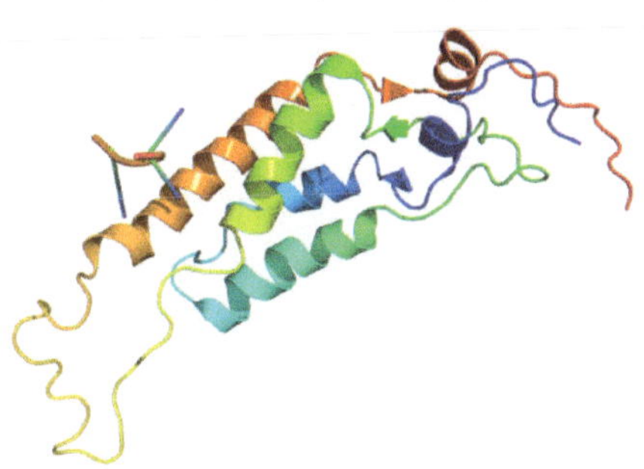	
A monomeric unit of the tobacco mosaic virus coat protein.	
Identifiers	
Symbol	CP
Entrez	1494073
UniProt	P03579

Tobacco mosaic virus has a rod-like appearance. Its capsid is made from 2130 molecules of coat protein and one molecule of genomic single strand RNA, 6400 bases long. The coat protein self-assembles into the rod-like helical structure (16.3 proteins per helix turn) around the RNA which forms a hairpin loop structure. The protein monomer consists of 158 amino acids which are assembled into four main alpha-helices, which are joined by a prominent loop proximal to the axis of the virion. Virions are ~300 nm in length and ~18 nm in diameter. Negatively stained electron microphotographs show a distinct inner channel of ~4 nm. The RNA is located at a radius of ~6 nm and is protected from the action of cellular enzymes by the coat protein. There are three RNA nucleotides per protein monomer. X-ray fiber diffraction structure of the intact virus was studied based on an electron density map at 3.6 Å resolution.

Genome

The TMV genome consists of a 6.3-6.5 kb single-stranded (ss) RNA. The 3'-terminus has a tRNA-like structure. The 5' terminus has a methylated nucleotide cap (m7G5'pppG). The genome encodes 4 open reading frames (ORFs), two of which produce a single protein due to ribosomal readthrough of a leaky UAG stop codon. The 4 genes encode a replicase (with methyltransferase [MT] and RNA helicase [Hel] domains), an RNA-dependent RNA polymerase, a so-called movement protein (MP) and a capsid protein (CP).

Physicochemical Properties

TMV is a thermostable virus. On a dried leaf, it can withstand up to 120 degrees Fahrenheit (50 °C) for 30 minutes.

TMV has an index of refraction of about 1.57.

Disease Cycle

TMV does not have a distinct overwintering structure. Rather, it will over-winter in infected tobacco stalks and leaves in the soil, on the surface of contaminated seed (TMV can even survive in contaminated tobacco products for many years). With the direct contact with host plants through its vectors (normally insects such as aphids and leafhoppers), TMV will go through the infection process and then the replication process.

Infection and Transmission

After its multiplication, it enters the neighboring cells through plasmodesmata. The infection spreads by direct contact to the neighboring cells, For its smooth entry, TMV produces a 30 kDa movement protein called P30 which enlarge the plasmodesmata. TMV most likely moves from cell-to-cell as a complex of the RNA, P30, and replicase proteins.

It can also spread through phloem for longer distance movement within the plant. Moreover, TMV can be transmitted from one plant to another by direct contact. Although TMV does not have defined transmission vectors, the virus can be easily transmitted from the infected hosts to the healthy plants, by human handling.

Replication

Following entry into its host via mechanical inoculation, TMV uncoats itself to release its viral [+] RNA strand. As uncoating occurs, the MetHel:Pol gene is translated to make the capping enzyme

MetHel and the RNA Polymerase. Then the viral genome will further replicate to produce multiple mRNAs via a [-]RNA intermediate primed by the $tRNA_{HIS}$ at the [+]RNA 3' end. The resulting mRNAs encode several proteins, including the coat protein and an RNA-dependent RNA polymerase (RdRp), as well as the movement protein. Thus TMV can replicate its own genome. After the coat protein and RNA genome of TMV have been synthesized, they spontaneously assemble into complete TMV virions in a highly organized process. The protomers come together to form disks or 'lockwashers' composed of two layers of protomers arranged in a helix. The helical capsid grows by the addition of protomers to the end of the rod. As the rod lengthens, the RNA passes through a channel in its center and forms a loop at the growing end. In this way the RNA can easily fit as a spiral into the interior of the helical capsid.

Host and Symptoms

Tobacco mosaic virus symptoms on tobacco

Tobacco mosaic virus symptoms on orchid

Like other plant pathogenic viruses, TMV has a very wide host range and has different effects depending on the host being infected. The tobacco mosaic virus has been known to cause a production loss for flue cured tobacco of up to two percent in North Carolina. It is known to infect members of nine plant families, and at least 125 individual species, including tobacco, tomato, pepper (all members of the useful Solanaceae), cucumbers, and a number of ornamental flowers. There are many different strains. The first symptom of this virus disease is a light green coloration between the veins of young leaves. This is followed quickly by the development of a "mosaic" or mottled pattern of light and dark green areas in the leaves. Rugosity may also be seen where the infected plant leaves display small localized random wrinkles. These symptoms develop quickly and are more pronounced on younger leaves. Its infection does not result in plant death, but if infection occurs early in the season, plants are stunted. Lower leaves are subjected to "mosaic burn" especially during periods of hot and dry weather. In these cases, large dead areas develop in the leaves. This constitutes one of the most destructive phases of tobacco mosaic virus infection. Infected leaves may be crinkled, puckered, or elongated. However, if TMV infects crops like grape and apple, it is almost symptomless.

Environment

TMV is known as one of the most stable viruses. It has a very wide survival range. As long as the surrounding temperature remains below approximately 40 degrees Celsius, TMV can sustain its stable form. All it needs is a host to infect. If necessary, greenhouses and botanical gardens would provide the most favorable condition for TMV to spread out, due to the high population density of possible hosts and the constant temperature throughout the year.

Treatment and Management

One of the common control methods for TMV is sanitation, which includes removing infected plants, and washing hands in between each planting. Crop rotation should also be employed to avoid infected soil/seed beds for at least two years. As for any plant disease, looking for resistant strains against TMV may also be advised. Furthermore, the cross protection method can be administered, where the stronger strain of TMV infection is inhibited by infecting the host plant with mild strain of TMV, similar to the effect of a vaccine.

In the past ten years, the application of genetic engineering on a host plant genome has been developed to allow the host plant to produce the TMV coat protein within their cells. It was hypothesized that the TMV genome will be re-coated rapidly upon entering the host cell, thus it prevents the initiation of TMV replication. Later it was found that the mechanism that protects the host from viral genome insertion is through gene silencing.

Scientific and Environmental Impact

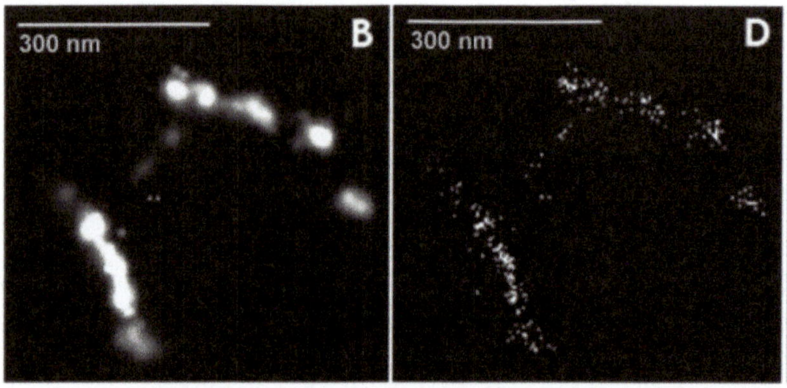

TMV virus: super resolution light microscopy

The large amount of literature about TMV and its choice for many pioneering investigations in structural biology (including X-ray diffraction), virus assembly and disassembly, and so on, are fundamentally due to the large quantities that can be obtained, plus the fact that it does not infect animals. After growing a few infected tobacco plants in a greenhouse and a few simple laboratory procedures, a scientist can easily produce several grams of the virus. As a result of this, TMV can be treated almost as an organic chemical, rather than an infective agent.

James D. Watson, in his memoir *The Double Helix*, cites his x-ray investigation of TMV's helical structure as an important step in deducing the nature of the DNA molecule.

Investigational uses

Due to its cylindrical shape, high aspect-ratio, self-assembling nature, and ability to incorporate metal coatings (nickel and cobalt) into its shell, TMV is an ideal candidate to be incorporated into battery electrodes Addition of TMV to a battery electrode increases the reactive surface area by an order of magnitude, resulting in an increase in the battery's capacity by up to six times compared to a planar electrode geometry.

Reverse Transcriptase

A reverse transcriptase (RT) is an enzyme used to generate complementary DNA (cDNA) from an RNA template, a process termed *reverse transcription*. It is mainly associated with retroviruses. However, non-retroviruses also use RT (for example, the hepatitis B virus, a member of the Hepadnaviridae, which are dsDNA-RT viruses, while retroviruses are ssRNA viruses). RT inhibitors are widely used as antiretroviral drugs. RT activities are also associated with the replication of chromosome ends (telomerase) and some mobile genetic elements (retrotransposons).

Retroviral RT has three sequential biochemical activities:

- RNA-dependent DNA polymerase activity,
- ribonuclease H, and
- DNA-dependent DNA polymerase activity.

These activities are used by the retrovirus to convert single-stranded genomic RNA into double-stranded cDNA which can integrate into the host genome, potentially generating a long-term infection that can be very difficult to eradicate. The same sequence of reactions is widely used in the laboratory to convert RNA to DNA for use in molecular cloning, RNA sequencing, polymerase chain reaction (PCR), or genome analysis.

Well studied reverse transcriptases include:

- HIV-1 reverse transcriptase from human immunodeficiency virus type 1 (PDB: 1HMV) has two subunits, which have respective molecular weights of 66 and 51 kDa.
- M-MLV reverse transcriptase from the Moloney murine leukemia virus is a single 75 kDa monomer.
- AMV reverse transcriptase from the avian myeloblastosis virus also has two subunits, a 63 kDa subunit and a 95 kDa subunit.
- Telomerase reverse transcriptase that maintains the telomeres of eukaryotic chromosomes.

History

Reverse transcriptases were discovered by Howard Temin at the University of Wisconsin–Madison in RSV virions and independently isolated by David Baltimore in 1970 at MIT from two RNA tumour viruses: R-MLV and again RSV. For their achievements, both shared the 1975 Nobel Prize in Physiology or Medicine (with Renato Dulbecco).

The idea of reverse transcription was very unpopular at first, as it contradicted the central dogma of molecular biology, which states that DNA is transcribed into RNA, which is then translated into proteins. However, in 1970, when the scientists Howard Temin and David Baltimore both independently discovered the enzyme responsible for reverse transcription, named reverse transcriptase, the possibility that genetic information could be passed on in this manner was finally accepted.

Function in Viruses

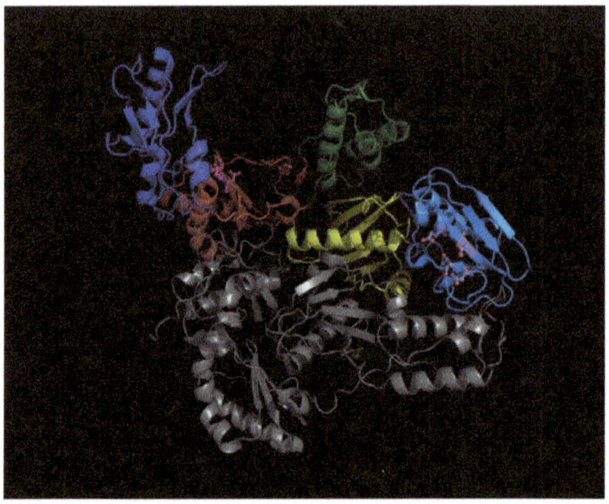

Reverse transcriptase is shown with its finger, palm, and thumb regions. The catalytic amino acids of the RNase H active site and the polymerase active site are shown in ball-and-stick form.

The enzymes are encoded and used by reverse-transcribing viruses, which use the enzyme during the process of replication. Reverse-transcribing RNA viruses, such as retroviruses, use the enzyme to reverse-transcribe their RNA genomes into DNA, which is then integrated into the host genome and replicated along with it. Reverse-transcribing DNA viruses, such as the hepadnaviruses, can allow RNA to serve as a template in assembling and making DNA strands. HIV infects humans with the use of this enzyme. Without reverse transcriptase, the viral genome would not be able to incorporate into the host cell, resulting in failure to replicate.

Process of Reverse Transcription

Reverse transcriptase creates single-stranded DNA from an RNA template.

In virus species with reverse transcriptase lacking DNA-dependent DNA polymerase activity, creation of double-stranded DNA can possibly be done by host-encoded DNA polymerase δ, mistaking the viral DNA-RNA for a primer and synthesizing a double-stranded DNA by similar mechanism as in primer removal, where the newly synthesized DNA displaces the original RNA template.

The process of reverse transcription is extremely error-prone, and it is during this step that mutations may occur. Such mutations may cause drug resistance.

Retroviral Reverse Transcription

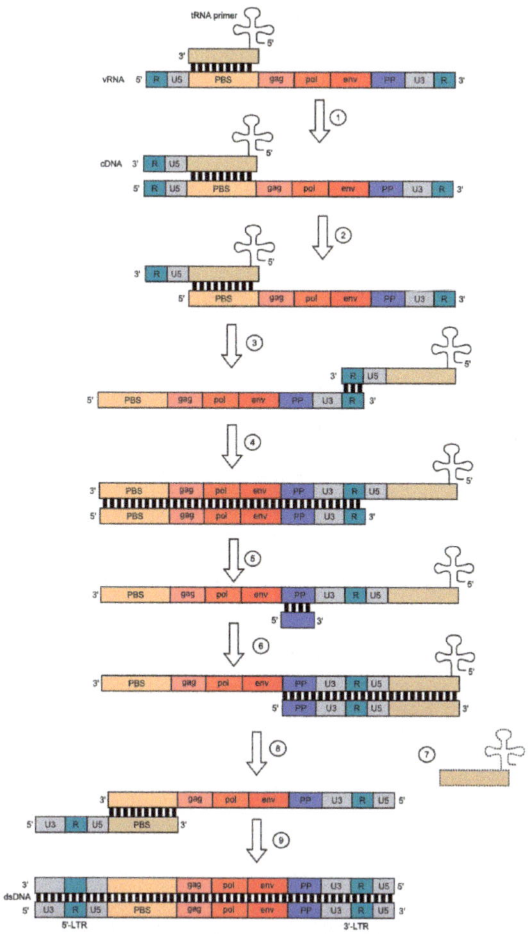

Retroviruses, also referred to as class VI ssRNA-RT viruses, are RNA reverse-transcribing viruses with a DNA intermediate. Their genomes consist of two molecules of positive-sense single-stranded RNA with a 5' cap and 3' polyadenylated tail. Examples of retroviruses include the human immunodeficiency virus (HIV) and the human T-lymphotropic virus (HTLV). Creation of double-stranded DNA occurs in the cytosol as a series of these steps:

1. A specific cellular tRNA acts as a primer and hybridizes to a complementary part of the virus RNA genome called the primer binding site or PBS.
2. Complementary DNA then binds to the U5 (non-coding region) and R region (a direct repeat found at both ends of the RNA molecule) of the viral RNA.
3. A domain on the reverse transcriptase enzyme called RNAse H degrades the 5' end of the RNA which removes the U5 and R region.
4. The primer then "jumps" to the 3' end of the viral genome, and the newly synthesised DNA strands hybridizes to the complementary R region on the RNA.
5. The first strand of complementary DNA (cDNA) is extended, and the majority of viral RNA is degraded by RNAse H.

6. Once the strand is completed, second strand synthesis is initiated from the viral RNA.
7. There is then another "jump" where the PBS from the second strand hybridizes with the complementary PBS on the first strand.
8. Both strands are extended further and can be incorporated into the hosts genome by the enzyme integrase.

Creation of double-stranded DNA also involves *strand transfer*, in which there is a translocation of short DNA product from initial RNA-dependent DNA synthesis to acceptor template regions at the other end of the genome, which are later reached and processed by the reverse transcriptase for its DNA-dependent DNA activity.

Retroviral RNA is arranged in 5' terminus to 3' terminus. The site where the primer is annealed to viral RNA is called the primer-binding site (PBS). The RNA 5'end to the PBS site is called U5, and the RNA 3' end to the PBS is called the leader. The tRNA primer is unwound between 14 and 22 nucleotides and forms a base-paired duplex with the viral RNA at PBS. The fact that the PBS is located near the 5' terminus of viral RNA is unusual because reverse transcriptase synthesize DNA from 3' end of the primer in the 5' to 3' direction (with respect to the RNA template). Therefore, the primer and reverse transcriptase must be relocated to 3' end of viral RNA. In order to accomplish this reposition, multiple steps and various enzymes including DNA polymerase, ribonuclease H(RNase H) and polynucleotide unwinding are needed.

The HIV reverse transcriptase also has ribonuclease activity that degrades the viral RNA during the synthesis of cDNA, as well as DNA-dependent DNA polymerase activity that copies the sense cDNA strand into an *antisense* DNA to form a double-stranded viral DNA intermediate (vDNA).

In Eukaryotes

Self-replicating stretches of eukaryotic genomes known as retrotransposons utilize reverse transcriptase to move from one position in the genome to another via an RNA intermediate. They are found abundantly in the genomes of plants and animals. Telomerase is another reverse transcriptase found in many eukaryotes, including humans, which carries its own RNA template; this RNA is used as a template for DNA replication.

In Prokaryotes

Initial reports of reverse transcriptase in prokaryotes came as far back as 1971 (Beljanski et al., 1971a, 1972). These have since been broadly described as part of bacterial Retrons, distinct sequences that code for reverse transcriptase, and are used in the synthesis of msDNA. In order to initiate synthesis of DNA, a primer is needed. In bacteria, the primer is synthesized during replication.

Evolutionary Role

Valerian Dolja of Oregon State argues that viruses due to their diversity have played an evolutionary role in the development of cellular life, with reverse transcriptase playing a central role.

Structure

Reverse transcriptase enzymes include an RNA-dependent DNA polymerase and a DNA-dependent DNA polymerase, which work together to perform transcription. In addition to the transcription function, retroviral reverse transcriptases have a domain belonging to the RNase H family, which is vital to their replication.

Replication Fidelity

There are three different replication systems during the life cycle of a retrovirus. First of all, the reverse transcriptase synthesizes viral DNA from viral RNA, and then from newly made complementary DNA strand. The second replication process occurs when host cellular DNA polymerase replicates the integrated viral DNA. Lastly, RNA polymerase II transcribes the proviral DNA into RNA, which will be packed into virions. Therefore, mutation can occur during one or all of these replication steps.

Reverse transcriptase has a high error rate when transcribing RNA into DNA since, unlike most other DNA polymerases, it has no proofreading ability. This high error rate allows mutations to accumulate at an accelerated rate relative to proofread forms of replication. The commercially available reverse transcriptases produced by Promega are quoted by their manuals as having error rates in the range of 1 in 17,000 bases for AMV and 1 in 30,000 bases for M-MLV.

Other than creating single-nucleotide polymorphisms, reverse transcriptases have also been shown to be involved in processes such as transcript fusions, exon shuffling and creating artificial antisense transcripts. It has been speculated that this *template switching* activity of reverse transcriptase, which can be demonstrated completely *in vivo*, may have been one of the causes for finding several thousand unannotated transcripts in the genomes of model organisms.

Applications

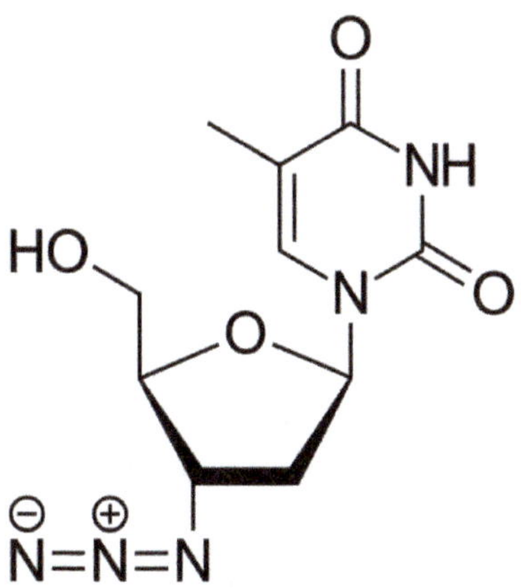

The molecular structure of zidovudine (AZT), a drug used to inhibit HIV reverse transcriptase

Antiviral Drugs

As HIV uses reverse transcriptase to copy its genetic material and generate new viruses (part of a retrovirus proliferation circle), specific drugs have been designed to disrupt the process and thereby suppress its growth. Collectively, these drugs are known as reverse-transcriptase inhibitors and include the nucleoside and nucleotide analogues zidovudine (trade name Retrovir), lamivudine (Epivir) and tenofovir (Viread), as well as non-nucleoside inhibitors, such as nevirapine (Viramune).

Molecular Biology

Reverse transcriptase is commonly used in research to apply the polymerase chain reaction technique to RNA in a technique called reverse transcription polymerase chain reaction (RT-PCR). The classical PCR technique can be applied only to DNA strands, but, with the help of reverse transcriptase, RNA can be transcribed into DNA, thus making PCR analysis of RNA molecules possible. Reverse transcriptase is used also to create cDNA libraries from mRNA. The commercial availability of reverse transcriptase greatly improved knowledge in the area of molecular biology, as, along with other enzymes, it allowed scientists to clone, sequence, and characterise RNA.

Reverse transcriptase has also been employed in insulin production. By inserting eukaryotic mRNA for insulin production along with reverse transcriptase into bacteria, the mRNA could be inserted into the prokaryote's genome. Large amounts of insulin can then be created, sidestepping the need to harvest pig pancreas and other such traditional sources. Directly inserting eukaryotic DNA into bacteria would not work because it carries introns, so would not translate successfully using the bacterial ribosomes. Processing in the eukaryotic cell during mRNA production removes these introns to provide a suitable template. Reverse transcriptase converted this edited RNA back into DNA so it could be incorporated in the genome.

Virus Shapes

Early study with tobacco mosaic virus (TMV) strongly suggested that viruses were composed of repeating subunits of protein which was later supported by crystallization of TMV. A major advancement in determining the morphology of virus was the development of negative stain electron microscopy . Another modification of classical electron microscopy is cryo-electron microscopy where the virus containing samples were rapidly frozen and examined at a very low temperature; this allows us to preserve the native structure of the viruses.

A virion is a complete virus particle that is surrounded by the capsid protein and encapsidates the viral genome (DNA or RNA). Sometime structure without nucleic acid can be visible under the electron microscope those structures are called as empty capsids . In some of the viruses like paramyxoviruses the nucleic acid is surrounded by the capsid proteins and the composite structures are referred as nucleocapsid . Some of the viruses contain the lipid envelope which surrounds the nucleocapsids. The envelopes are derived from the host cell membrane during the budding process. As the envelopes are derived from host cell membrane they contain many of the surface proteins present in the host cells.

There are two kinds of symmetry found among the viruses: icosahedral and helical . In theory the icosahedral symmetry may sometime referred as spherical based on the external morphology.

Icosahedral symmetry has 12 vertices , 30 edges , and 20 faces . They also have two, three, or five fold symmetry based on the rotation through axes passing through their edges, faces, and vertices respectively. The viruses of this kind look spherical in shape. In helical symmetry the genomic RNA forms a spiral within the core of the nucleocapsids. The viruses of this kind look rodlike or filamentous . The viruses which contain large DNA genomes are more complex in structure, for example- poxviruses and herpesviruses.

An icosahedral virion structure showing two, three, and fivefold symmetry.

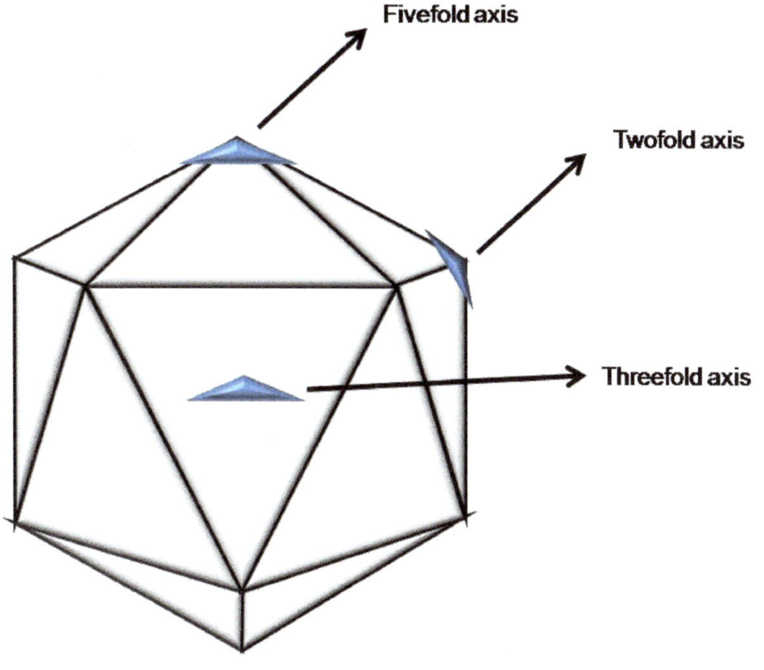

Virus structure with helical symmetry.

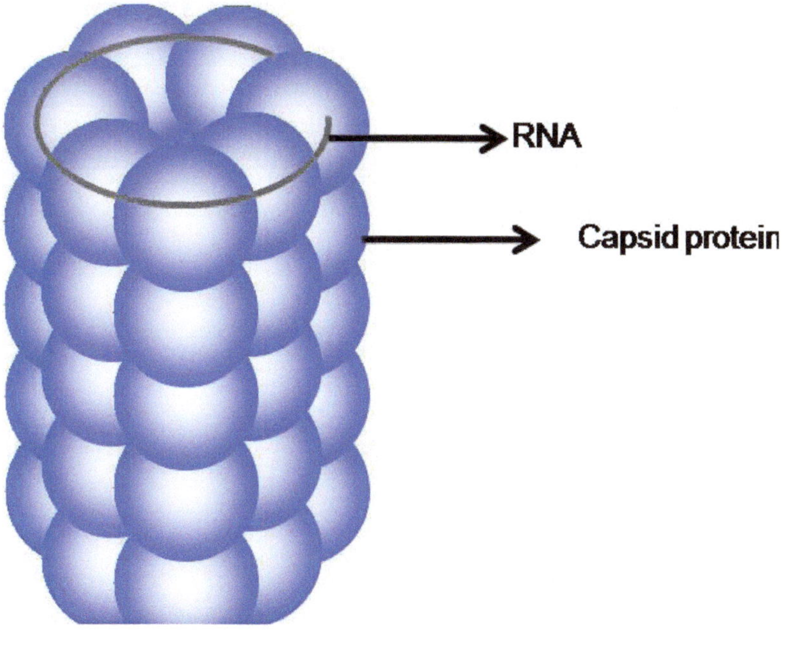

Shape of viruses belonging to different families:

Family	Shape
Poxviridae	Pleomorphic
Iridoviridae	Icosahedral
Asfarviridae	Spherical
Herpesviridae	Icosahedral
Adenoviridae	Icosahedral
Polyomaviridae	Icosahedral
Papillomaviridae	Icosahedral
Hepadnaviridae	Spherical
Circoviridae	Icosahedral
Parvoviridae	Icosahedral
Retroviridae	Spherical
Reoviridae	Icosahedral
Birnaviridae	Icosahedral
Paramyxoviridae	Pleomorphic
Rhabdoviridae	Bullet shaped
Filoviridae	Filamentous
Bornaviridae	Spherical
Orthomyxoviridae	Pleomorphic
Bunyaviridae	Spherical
Arenaviridae	Spherical
Coronaviridae	Spherical
Arteriviridae	Spherical
Picornaviridae	Icosahedral
Caliciviridae	Icosahedral
Astroviridae	Icosahedral
Togaviridae	Spherical
Flaviviridae	Spherical

Capsid

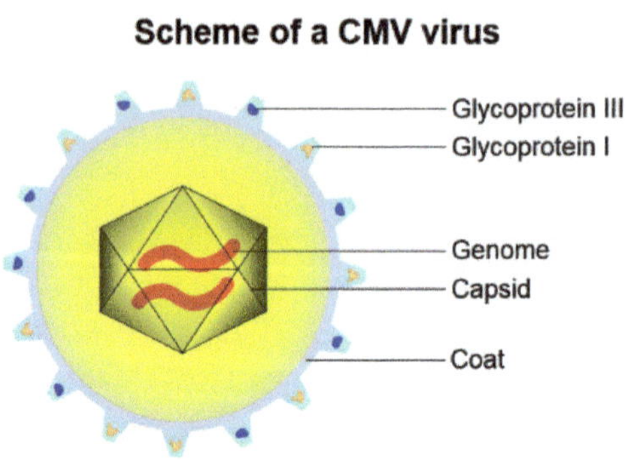

Schematic of a cytomegalovirus

A capsid is the protein shell of a virus. It consists of several oligomeric structural subunits made of protein called protomers. The observable 3-dimensional morphological subunits, which may or may not correspond to individual proteins, are called capsomeres. The capsid encloses the genetic material of the virus.

Capsids are broadly classified according to their structure. The majority of viruses have capsids with either helical or icosahedral structure. Some viruses, such as bacteriophages, have developed more complicated structures due to constraints of elasticity and electrostatics. The icosahedral shape, which has 20 equilateral triangular faces, approximates a sphere, while the helical shape resembles the shape of a spring, taking the space of a cylinder but not being a cylinder itself. The capsid faces may consist of one or more proteins. For example, the foot-and-mouth disease virus capsid has faces consisting of three proteins named VP1–3.

Some viruses are *enveloped*, meaning that the capsid is coated with a lipid membrane known as the *viral envelope*. The envelope is acquired by the capsid from an intracellular membrane in the virus' host; examples include the inner nuclear membrane, the golgi membrane, and the cell's outer membrane.

Once the virus has infected a cell and begins replicating itself, new capsid subunits are synthesized according to the genetic material of the virus, using the protein biosynthesis mechanism of the cell. During the assembly process, a portal subunit is assembled at one vertex of the capsid. Through this portal, viral DNA or RNA is transported into the capsid.

Structural analyses of major capsid protein (MCP) architectures have been used to categorise viruses into families. For example, the bacteriophage PRD1, Paramecium bursaria Chlorella algal virus, and mammalian adenovirus have been placed in the same family.

Specific Shapes

Icosahedral

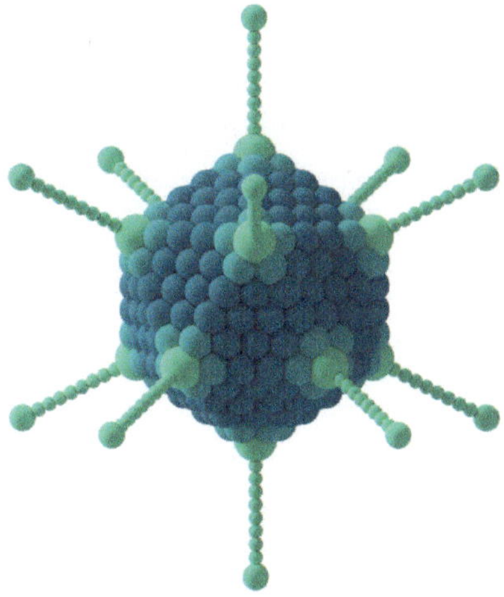

Icosahedral capsid of an adenovirus

Although the icosahedral structure is extremely common among viruses, size differences and slight variations exist between virions. Given an asymmetric subunit on a triangular face of a regular icosahedron, with three subunits per face 60 such subunits can be placed in an equivalent manner. Most virions, because of their size, have more than 60 subunits. These variations have been classified on the basis of the quasi-equivalence principle proposed by Donald Caspar and Aaron Klug.

An icosahedral structure can be regarded as being constructed from 12 pentamers. The number of pentamers is fixed but the number of hexamers can vary. These shells can be constructed from pentamers and hexamers by minimizing the number T (triangulation number) of nonequivalent locations that subunits occupy, with the T-number adopting the particular integer values 1, 3, 4, 7, 12, 13,...($T = h^2 + k^2 + hk$, with h, k equal to nonnegative integers). These shells always contain 12 pentamers plus 10 (T-1) hexamers. Although this classification can be applied to the majority of known viruses exceptions are known including the retroviruses where point mutations disrupt the symmetry.

Prolate

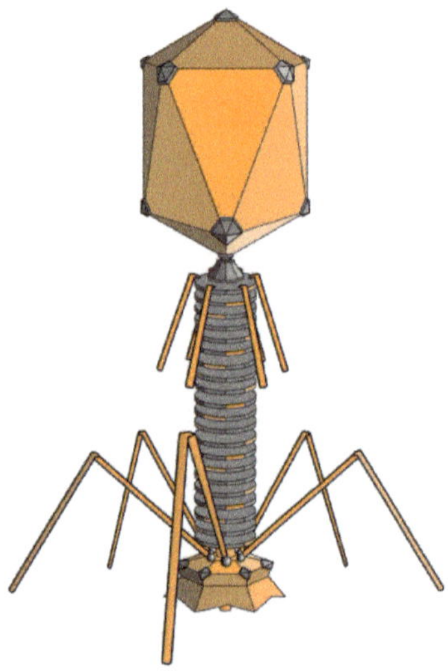

The prolate structure of a typical head on a bacteriophage

This is an icosahedron elongated along the fivefold axis and is a common arrangement of the heads of bacteriophages. Such a structure is composed of a cylinder with a cap at either end. The cylinder is composed of 10 triangles. The Q number, which can be any positive integer, specifies the number of triangles, composed of asymmetric subunits, that make up the 10 triangles of the cylinder. The caps are classified by the T number. An elongated, stretched-out, shperical capsid is characterisitc of the prolate shape.

Helical

Many rod-shaped and filamentous plant viruses have capsids with helical symmetry. The helical structure can be described as a set of n 1-D molecular helices related by an n-fold axial symmetry. The helical transformation are classified into two categories: one-dimensional and two-dimensional helical systems. Creating an entire helical structure relies on a set of translational and rotational matrices which are coded in the protein data bank. Helical symmetry is given by the formula $P=\mu \times \rho$, where μ is the number of structural units per turn of the helix, ρ is the axial rise per unit and P is the pitch of the helix. The structure is said to be open due to the characteristic that any volume can be enclosed by varying the length of the helix. The most understood helical virus is the tobacco mosaic virus. The virus is a single molecule of (+) strand RNA. Each coat protein on the interior of the helix bind three nucleotides of the RNA genome. Influenza A viruses differ by comprising multiple ribonucleoproteins, the viral NP protein organizes the RNA into a helical structure. The size is also different the tobacco mosaic virus has a 16.33 protein subunits per helical turn, while the influenza A virus has a 28 amino acid tail loop.

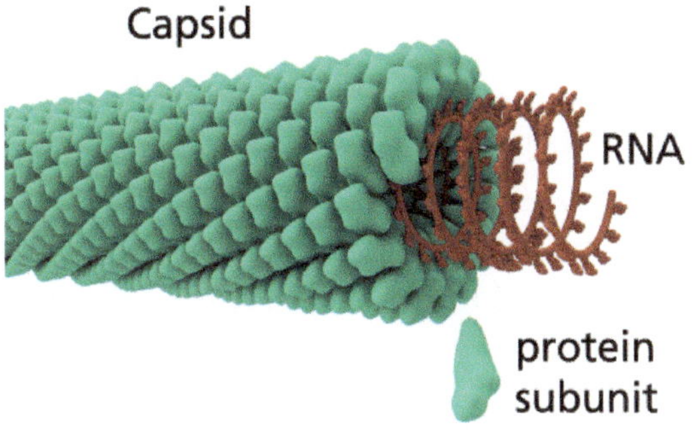

3D model of a helical capsid structure of a virus

Triangulation Number

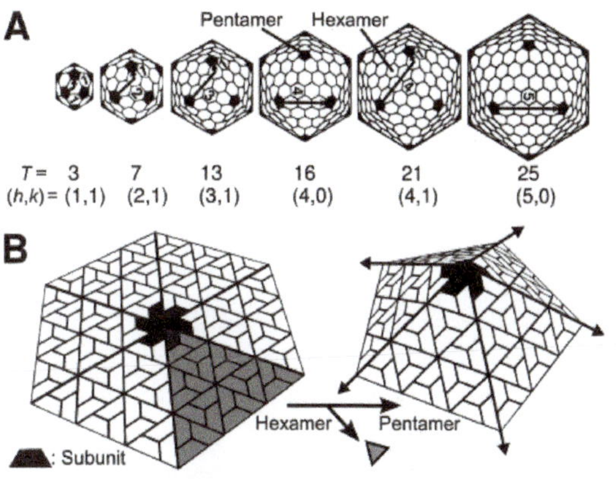

Virus capsid T-numbers

Icosahedral virus capsids are typically assigned a triangulation number (T-number) to describe the relation between the number of pentagons and hexagons *i.e.* their quasi-symmetry in the capsid shell. The T-number idea was originally developed to explain the quasi-symmetry by Caspar and Klug in 1962.

For example, a purely dodecahedral virus has a T-number of 1 (usually written, T=1) and a truncated icosahedron is assigned T=3. The T-number is calculated by (1) applying a grid to the surface of the virus with coordinates *h* and *k*, (2) counting the number of steps between successive pentagons on the virus surface, (3) applying the formula:

$$\mathrm{T} = \mathrm{h}^2 + \mathrm{h} \cdot \mathrm{k} + \mathrm{k}^2 = (h+k)^2 - hk$$

where $h \geq k$ and h and k are the distances between the successive pentagons on the virus surface for each axis. The larger the T-number the more hexagons are present relative to the pentagons.

For the hexagonal system, the polyhedra have $20T$ vertices, $30T$ edges, $10T+2$ faces (12 pentagons and $10(T-1)$ hexagons). For the dual triangular, the vertex and face counts are flipped.

T-numbers can be represented in different ways, for example T=1 can only be represented as an icosahedron or a dodecahedron and, depending on the type of quasi-symmetry, T= 3 can be presented as a truncated dodecahedron, an icosidodecahedron, or a truncated icosahedron and their respective duals a triakis icosahedron, a rhombic triacontahedron, or a pentakis dodecahedron.

Functions

The functions of the virion are to protect the genome, deliver the genome and interact with the host. The virion must assemble a stable, protective protein shell to protect the genome from lethal chemical and physical agents. These include forms of natural radiation, extremes of pH or temperature and proteolytic and nucleolytic enzymes. Delivery of the genome is also important by specific binding to external receptors of the host cell, transmission of specific signals that induce uncoating of the genome, and induction of fusion with host cell membranes.

Chemical properties

The viral particle must be metastable so that interactions can be reversed readily when entering and uncoating a new host cell. If it attains the minimum free energy state conformation will be irreversible associated with attachment and entry. Each subunit of the capsid has identical bonding contacts with its neighbors, and the two binding contacts are usually noncovalent. The non-covalent bonding holds the structural unit together. The reversible formation of non-covalent bonds between properly folded subunits leads to error-free assembly and minimizes free energy.

Transmission Electron Microscopy

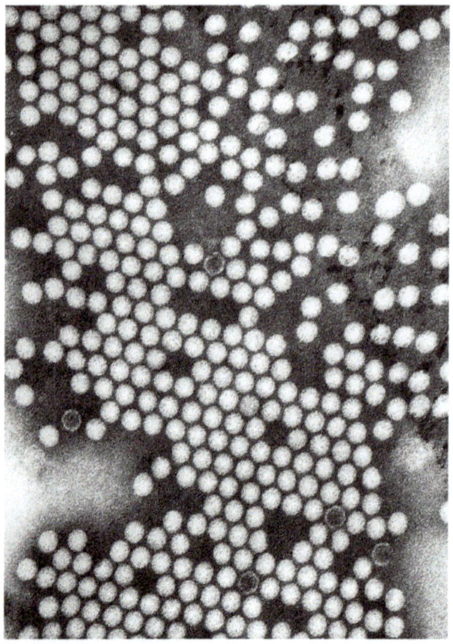

A TEM image of the polio virus. The polio virus is 30 nm in size.

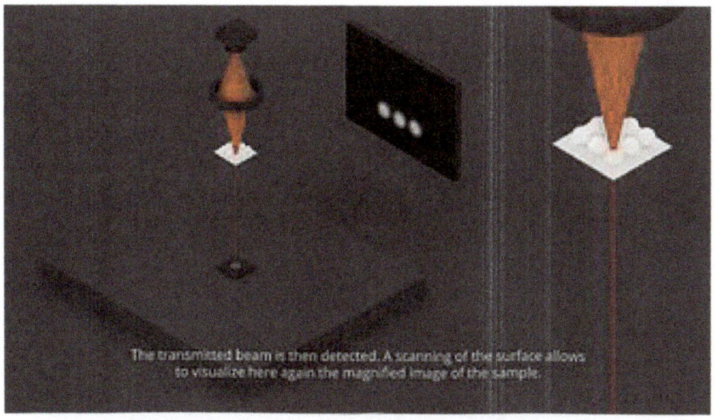

Operating principle of a Transmission Electron Microscope

Transmission electron microscopy (TEM) is a microscopy technique in which a beam of electrons is transmitted through an ultra-thin specimen, interacting with the specimen as it passes through it. An image is formed from the interaction of the electrons transmitted through the specimen; the image is magnified and focused onto an imaging device, such as a fluorescent screen, on a layer of photographic film, or to be detected by a sensor such as a charge-coupled device.

TEMs are capable of imaging at a significantly higher resolution than light microscopes, owing to the small de Broglie wavelength of electrons. This enables the instrument's user to examine fine detail—even as small as a single column of atoms, which is thousands of times smaller than the smallest resolvable object in a light microscope. TEM forms a major analysis method in a range of scientific fields, in physical, chemical and biological sciences. TEMs find application in cancer research, virology, materials science as well as pollution, nanotechnology and semiconductor research.

At smaller magnifications TEM image contrast is due to absorption of electrons in the material, due to the thickness and composition of the material. At higher magnifications complex wave interactions modulate the intensity of the image, requiring expert analysis of observed images. Alternate modes of use allow for the TEM to observe modulations in chemical identity, crystal orientation, electronic structure and sample induced electron phase shift as well as the regular absorption based imaging.

The first TEM was built by Max Knoll and Ernst Ruska in 1931, with this group developing the first TEM with resolution greater than that of light in 1933 and the first commercial TEM in 1939. In 1986, Ruska was awarded the Nobel Prize in physics for the development of transmission electron microscopy.

History

Initial Development

In 1873, Ernst Abbe proposed that the ability to resolve detail in an object was limited approximately by the wavelength of the light used in imaging, which limits the resolution of an optical microscope to a few hundred nanometers. Developments into ultraviolet (UV) microscopes, led by Köhler and Rohr, allowed for an increase in resolving power of about a factor of two. However this required more expensive quartz optical components, due to the absorption of UV by glass. At this point it was believed that obtaining an image with sub-micrometer information was simply impossible due to this wavelength constraint.

The first practical TEM, originally installed at IG Farben-Werke and now on display at the Deutsches Museum in Munich, Germany

It had earlier been recognized by Plücker in 1858 that the deflection of "cathode rays" (electrons) was possible by the use of magnetic fields. This effect had been utilized to build primitive cathode ray oscilloscopes (CROs) as early as 1897 by Ferdinand Braun, intended as a measurement device. Indeed, in 1891 it was recognized by Riecke that the cathode rays could be focused by these magnetic fields, allowing for simple lens designs. Later this theory was extended by Hans Busch in his work published in 1926, who showed that the lens maker's equation could, under appropriate assumptions, be applicable to electrons.

In 1928, at the Technological University of Berlin Adolf Matthias, Professor of High voltage Technology and Electrical Installations, appointed Max Knoll to lead a team of researchers to advance the CRO design. The team consisted of several PhD students including Ernst Ruska and Bodo von Borries. This team of researchers concerned themselves with lens design and CRO column placement, which they attempted to obtain the parameters that could be optimized to allow for construction of better CROs, as well as the development of electron optical components which could be used to generate low magnification (nearly 1:1) images. In 1931 the group successfully generated magnified images of mesh grids placed over the anode aperture. The device used two magnetic lenses to achieve higher magnifications, arguably the first electron microscope. In that same year, Reinhold Rudenberg, the scientific director of the Siemens company, had patented an electrostatic lens electron microscope.

Improving Resolution

At this time the wave nature of electrons, which were considered charged matter particles, had not been fully realized until the publication of the De Broglie hypothesis in 1927. The group was unaware of this publication until 1932, where it was quickly realized that the De Broglie wavelength of electrons was many orders of magnitude smaller than that for light, theoretically allowing for imaging at atomic scales. In April 1932, Ruska suggested the construction of a new electron microscope for direct imaging of specimens inserted into the microscope, rather than simple mesh grids or images of apertures. With this device successful diffraction and normal imaging of aluminium sheet was achieved, however exceeding the magnification achievable with light microscopy had still not been successfully demonstrated. This goal was achieved in September 1933, using images of cotton fibers, which were quickly acquired before being damaged by the electron beam.

At this time, interest in the electron microscope had increased, with other groups, such as Paul Anderson and Kenneth Fitzsimmons of Washington State University, and Albert Prebus and James Hillier at the University of Toronto who constructed the first TEMs in North America in 1935 and 1938, respectively, continually advancing TEM design.

Research continued on the electron microscope at Siemens in 1936, the aim of the research was the development improvement of TEM imaging properties, particularly with regard to biological specimens. At this time electron microscopes were being fabricated for specific groups, such as the "EM1" device used at the UK National Physical Laboratory. In 1939 the first commercial electron microscope, pictured, was installed in the Physics department of IG Farben-Werke. Further work on the electron microscope was hampered by the destruction of a new laboratory constructed at Siemens by an air-raid, as well as the death of two of the researchers, Heinz Müller and Friedrick Krause during World War II.

Further Research

After World War II, Ruska resumed work at Siemens, where he continued to develop the electron microscope, producing the first microscope with 100k magnification. The fundamental structure of this microscope design, with multi-stage beam preparation optics, is still used in modern microscopes. The worldwide electron microscopy community advanced with electron microscopes being manufactured in Manchester UK, the USA (RCA), Germany (Siemens) and Japan (JEOL). The first international conference in electron microscopy was in Delft in 1949, with more than one hundred attendees. Later conferences included the "First" international conference in Paris, 1950 and then in London in 1954.

With the development of TEM, the associated technique of scanning transmission electron microscopy (STEM) was re-investigated and did not become developed until the 1970s, with Albert Crewe at the University of Chicago developing the field emission gun and adding a high quality objective lens to create the modern STEM. Using this design, Crewe demonstrated the ability to image atoms using annular dark-field imaging. Crewe and coworkers at the University of Chicago developed the cold field electron emission source and built a STEM able to visualize single heavy atoms on thin carbon substrates. In 2008, Jannick Meyer et al. described the direct visualization of light atoms such as carbon and even hydrogen using TEM and a clean single-layer graphene substrate.

Electrons

Theoretically, the maximum resolution, *d*, that one can obtain with a light microscope has been limited by the wavelength of the photons that are being used to probe the sample, λ and the numerical aperture of the system, *NA*.

$$d = \frac{\lambda}{2n\sin\alpha} \approx \frac{\lambda}{2NA}$$

where n is the index of refraction of the medium in which the lens is working and α is the angle of light that can enter the lens. Early twentieth century scientists theorized ways of getting around the limitations of the relatively large wavelength of visible light (wavelengths of 400–700 nanometers) by using electrons. Like all matter, electrons have both wave and particle properties (as theorized by Louis-Victor de Broglie), and their wave-like properties mean that a beam of electrons can be made to behave like a beam of electromagnetic radiation. The wavelength of electrons is related to their kinetic energy via the de Broglie equation. An additional correction must be made to account for relativistic effects, as in a TEM an electron's velocity approaches the speed of light, *c*.

$$\lambda_e \approx \frac{h}{\sqrt{2m_0E\left(1+\frac{E}{2m_0c^2}\right)}}$$

where, *h* is Planck's constant, m_0 is the rest mass of an electron and *E* is the energy of the accelerated electron. Electrons are usually generated in an electron microscope by a process known as thermionic emission from a filament, usually tungsten, in the same manner as a light bulb, or alternatively by field electron emission. The electrons are then accelerated by an electric potential (measured in volts) and focused by electrostatic and electromagnetic lenses onto the sample. The transmitted beam contains information about electron density, phase and periodicity; this beam is used to form an image.

Source Formation

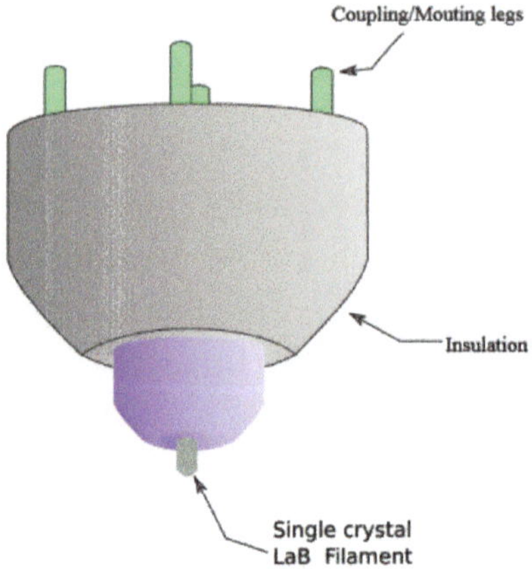

Single crystal LaB_6 filament

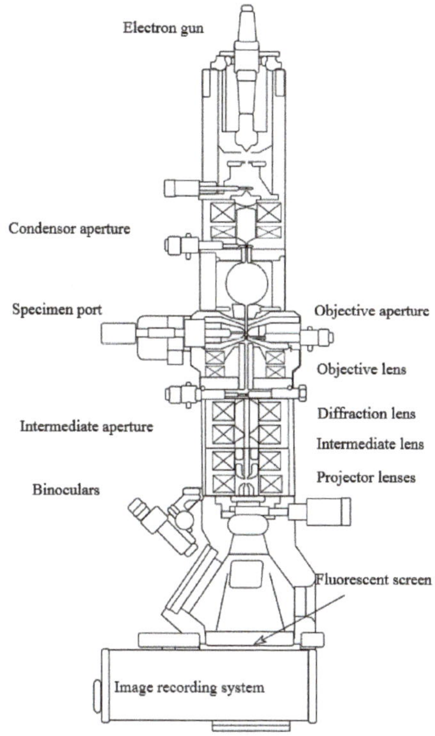

Layout of optical components in a basic TEM

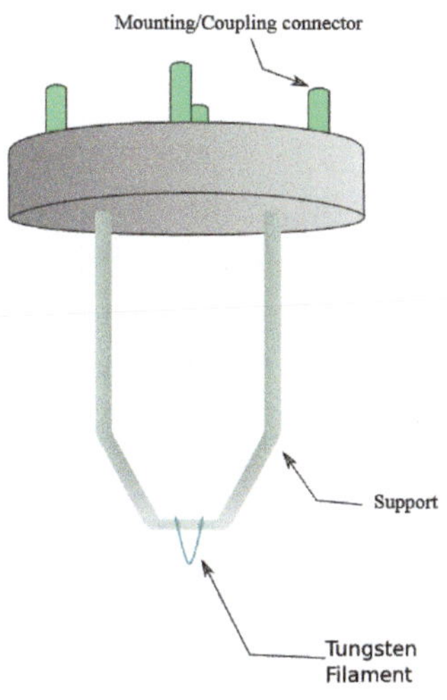

Hairpin style tungsten filament

From the top down, the TEM consists of an emission source, which may be a tungsten filament, or a lanthanum hexaboride (LaB_6) source. For tungsten, this will be of the form of either a hairpin-style filament, or a small spike-shaped filament. LaB_6 sources utilize small single crystals. By connecting this gun to a high voltage source (typically ~100–300 kV) the gun will, given sufficient current, begin to emit electrons either by thermionic or field electron emission into the vacuum. This extraction is almost always aided by the use of a Wehnelt cylinder to provide preliminary focus by consolidating and directing the electrons in these initial phases of forming the emitted electrons into a beam. The upper lenses of the TEM then further focus the electron beam to the desired size and location for subsequent interaction with the sample.

Manipulation of the electron beam is performed using two physical effects. The interaction of electrons with a magnetic field will cause electrons to move according to the left hand rule, thus allowing for electromagnets to manipulate the electron beam. The use of magnetic fields allows for the formation of a magnetic lens of variable focusing power, the lens shape originating due to the distribution of magnetic flux. Additionally, electrostatic fields can cause the electrons to be deflected through a constant angle. Coupling of two deflections in opposing directions with a small intermediate gap allows for the formation of a shift in the beam path, this being used in TEM for beam shifting, subsequently this is extremely important to STEM. From these two effects, as well as the use of an electron imaging system, sufficient control over the beam path is possible for TEM operation. The optical configuration of a TEM can be rapidly changed, unlike that for an optical microscope, as lenses in the beam path can be enabled, have their strength changed, or be disabled entirely simply via rapid electrical switching, the speed of which is limited by effects such as the magnetic hysteresis of the lenses.

Optics

The lenses of a TEM allow for beam convergence, with the angle of convergence as a variable parameter, giving the TEM the ability to change magnification simply by modifying the amount of current that flows through the coil, quadrupole or hexapole lenses. The quadrupole lens is an arrangement of electromagnetic coils at the vertices of the square, enabling the generation of a lensing magnetic fields, the hexapole configuration simply enhances the lens symmetry by using six, rather than four coils.

Typically a TEM consists of three stages of lensing. The stages are the condenser lenses, the objective lenses, and the projector lenses. The condenser lenses are responsible for primary beam formation, while the objective lenses focus the beam that comes through the sample itself (in STEM scanning mode, there are also objective lenses above the sample to make the incident electron beam convergent). The projector lenses are used to expand the beam onto the phosphor screen or other imaging device, such as film. The magnification of the TEM is due to the ratio of the distances between the specimen and the objective lens' image plane. Additional stigmators allow for the correction of asymmetrical beam distortions, known as astigmatism. It is noted that TEM optical configurations differ significantly with implementation, with manufacturers using custom lens configurations, such as in spherical aberration corrected instruments, or TEMs utilizing energy filtering to correct electron chromatic aberration.

Display

Imaging systems in a TEM consist of a phosphor screen, which may be made of fine (10–100 μm) particulate zinc sulfide, for direct observation by the operator, and, optionally, an image recording system such as film based or doped YAG screen coupled CCDs. Typically these devices can be removed or inserted into the beam path by the operator as required.

Components

A TEM is composed of several components, which include a vacuum system in which the electrons travel, an electron emission source for generation of the electron stream, a series of electromagnetic lenses, as well as electrostatic plates. The latter two allow the operator to guide and manipulate the beam as required. Also required is a device to allow the insertion into, motion within, and removal of specimens from the beam path. Imaging devices are subsequently used to create an image from the electrons that exit the system.

The electron source of the TEM is at the top, where the lensing system (4,7 and 8) focuses the beam on the specimen and then projects it onto the viewing screen (10). The beam control is on the right (13 and 14)

Vacuum System

To increase the mean free path of the electron gas interaction, a standard TEM is evacuated to low pressures, typically on the order of 10^{-4} Pa. The need for this is twofold: first the allowance for the voltage difference between the cathode and the ground without generating an arc, and secondly to reduce the collision frequency of electrons with gas atoms to negligible levels—this effect is characterized by the mean free path. TEM components such as specimen holders and film cartridges must be routinely inserted or replaced requiring a system with the ability to re-evacuate on a regular basis. As such, TEMs are equipped with multiple pumping systems and airlocks and are not permanently vacuum sealed.

The vacuum system for evacuating a TEM to an operating pressure level consists of several stages. Initially, a low or roughing vacuum is achieved with either a rotary vane pump or diaphragm pumps setting a sufficiently low pressure to allow the operation of a turbo-molecular or diffusion pump establishing high vacuum level necessary for operations. To allow for the low vacuum pump to not require continuous operation, while continually operating the turbo-molecular pumps, the vacuum side of a low-pressure pump may be connected to chambers which accommodate the exhaust gases from the turbo-molecular pump. Sections of the TEM may be isolated by the use of pressure-limiting apertures to allow for different vacuum levels in specific areas such as a higher vacuum of 10^{-4} to 10^{-7} Pa or higher in the electron gun in high-resolution or field-emission TEMs.

High-voltage TEMs require ultra-high vacuums on the range of 10^{-7} to 10^{-9} Pa to prevent the generation of an electrical arc, particularly at the TEM cathode. As such for higher voltage TEMs a third vacuum system may operate, with the gun isolated from the main chamber either by gate valves or a differential pumping aperture – a small hole that prevents the diffusion of gas molecules into the higher vacuum gun area faster than they can be pumped out. For these very low pressures, either an ion pump or a getter material is used.

Poor vacuum in a TEM can cause several problems ranging from the deposition of gas inside the TEM onto the specimen while viewed in a process known as electron beam induced deposition to more severe cathode damages caused by electrical discharge. The use of a cold trap to adsorb sublimated gases in the vicinity of the specimen largely eliminates vacuum problems that are caused by specimen sublimation.

Specimen Stage

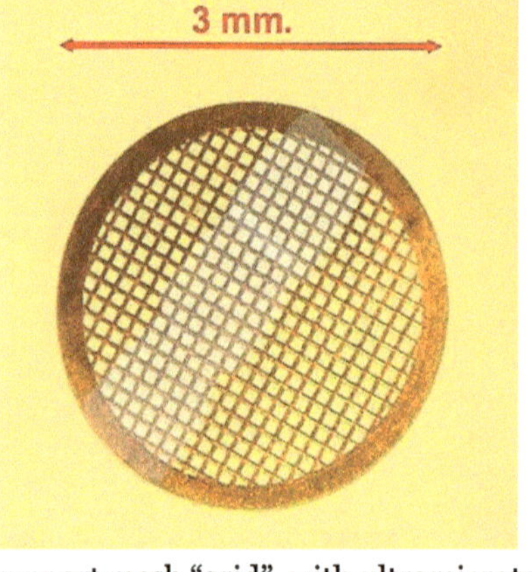

TEM sample support mesh "grid", with ultramicrotomy sections

TEM specimen stage designs include airlocks to allow for insertion of the specimen holder into the vacuum with minimal increase in pressure in other areas of the microscope. The specimen holders are adapted to hold a standard size of grid upon which the sample is placed or a standard size of self-supporting specimen. Standard TEM grid sizes are a 3.05 mm diameter ring, with a thickness and mesh size ranging from a few to 100 μm. The sample is placed onto the inner meshed area having diameter of approximately 2.5 mm. Usual grid materials are copper, molybdenum, gold or platinum. This grid is placed into the sample holder, which is paired with the specimen stage. A wide variety of designs of stages and holders exist, depending upon the type of experiment being performed. In addition to 3.05 mm grids, 2.3 mm grids are sometimes, if rarely, used. These grids were particularly used in the mineral sciences where a large degree of tilt can be required and where specimen material may be extremely rare. Electron transparent specimens have a thickness around 100 nm, but this value depends on the accelerating voltage.

Once inserted into a TEM, the sample often has to be manipulated to present the region of interest to the beam, such as in single grain diffraction, in a specific orientation. To accommodate this, the TEM stage includes mechanisms for the translation of the sample in the XY plane of the sample, for Z height adjustment of the sample holder, and usually for at least one rotation degree of freedom for the sample. Thus a TEM stage may provide four degrees of freedom for the motion of the specimen. Most modern TEMs provide the ability for two orthogonal rotation angles of movement with specialized holder designs called double-tilt sample holders. Of note however is that some stage designs, such as top-entry or vertical insertion stages once common for high resolution TEM studies, may simply only have X-Y translation available. The design criteria of TEM stages are complex, owing to the simultaneous requirements of mechanical and electron-optical constraints and have thus generated many unique implementations.

A TEM stage is required to have the ability to hold a specimen and be manipulated to bring the region of interest into the path of the electron beam. As the TEM can operate over a wide range of magnifications, the stage must simultaneously be highly resistant to mechanical drift, with drift requirements as low as a few nm/minute while being able to move several μm/minute, with repositioning accuracy on the order of nanometers. Earlier designs of TEM accomplished this with a complex set of mechanical downgearing devices, allowing the operator to finely control the motion of the stage by several rotating rods. Modern devices may use electrical stage designs, using screw gearing in concert with stepper motors, providing the operator with a computer-based stage input, such as a joystick or trackball.

Two main designs for stages in a TEM exist, the side-entry and top entry version. Each design must accommodate the matching holder to allow for specimen insertion without either damaging delicate TEM optics or allowing gas into TEM systems under vacuum.

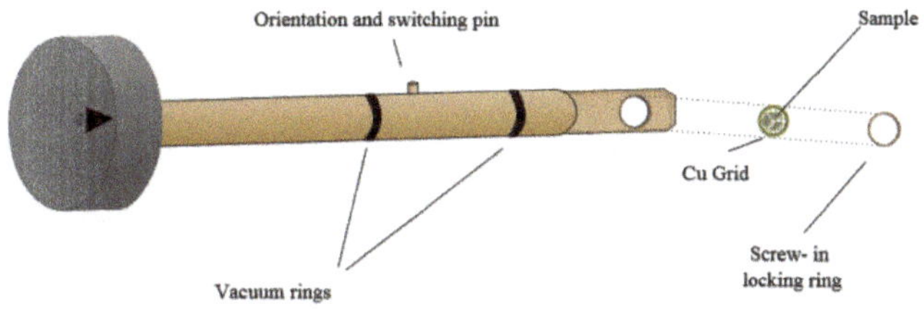

A diagram of a single axis tilt sample holder for insertion into a TEM goniometer. Titling of the holder is achieved by rotation of the entire goniometer

The most common is the side entry holder, where the specimen is placed near the tip of a long metal (brass or stainless steel) rod, with the specimen placed flat in a small bore. Along the rod are several polymer vacuum rings to allow for the formation of a vacuum seal of sufficient quality, when inserted into the stage. The stage is thus designed to accommodate the rod, placing the sample either in between or near the objective lens, dependent upon the objective design. When inserted into the stage, the side entry holder has its tip contained within the TEM vacuum, and the base is presented to atmosphere, the airlock formed by the vacuum rings.

Insertion procedures for side-entry TEM holders typically involve the rotation of the sample to trigger micro switches that initiate evacuation of the airlock before the sample is inserted into the TEM column.

The second design is the top-entry holder consists of a cartridge that is several cm long with a bore drilled down the cartridge axis. The specimen is loaded into the bore, possibly utilizing a small screw ring to hold the sample in place. This cartridge is inserted into an airlock with the bore perpendicular to the TEM optic axis. When sealed, the airlock is manipulated to push the cartridge such that the cartridge falls into place, where the bore hole becomes aligned with the beam axis, such that the beam travels down the cartridge bore and into the specimen. Such designs are typically unable to be tilted without blocking the beam path or interfering with the objective lens.

Electron Gun

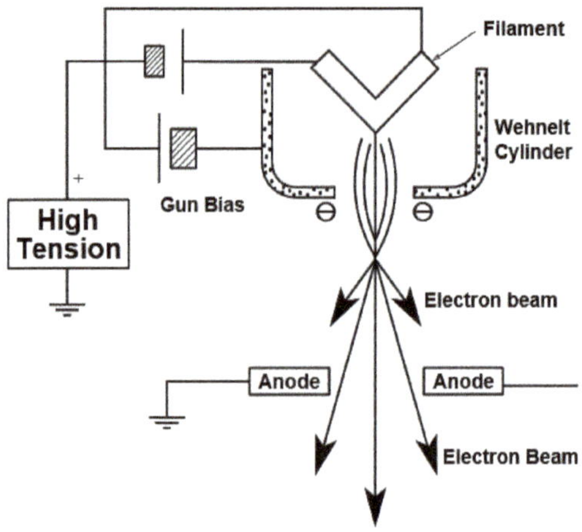

Cross sectional diagram of an electron gun assembly, illustrating electron extraction

The electron gun is formed from several components: the filament, a biasing circuit, a Wehnelt cap, and an extraction anode. By connecting the filament to the negative component power supply, electrons can be "pumped" from the electron gun to the anode plate, and TEM column, thus completing the circuit. The gun is designed to create a beam of electrons exiting from the assembly at some given angle, known as the gun divergence semi-angle, α. By constructing the Wehnelt cylinder such that it has a higher negative charge than the filament itself, electrons that exit the filament in a diverging manner are, under proper operation, forced into a converging pattern the minimum size of which is the gun crossover diameter.

The thermionic emission current density, *J*, can be related to the work function of the emitting material via Richardson's law

$$J = AT^2 \exp\left(-\frac{\Phi}{kT}\right),$$

where *A* is the Richardson's constant, Φ is the work function and T is the temperature of the material.

This equation shows that in order to achieve sufficient current density it is necessary to heat the emitter, taking care not to cause damage by application of excessive heat, for this reason materials with either a high melting point, such as tungsten, or those with a low work function (LaB_6) are required for the gun filament. Furthermore, both lanthanum hexaboride and tungsten thermionic sources must be heated in order to achieve thermionic emission, this can be achieved by the use of a small resistive strip. To prevent thermal shock, there is often a delay enforced in the application of current to the tip, to prevent thermal gradients from damaging the filament, the delay is usually a few seconds for LaB_6, and significantly lower for tungsten.

Electron Lens

Electron lenses are designed to act in a manner emulating that of an optical lens, by focusing parallel electrons at some constant focal distance. Electron lenses may operate electrostatically or magnetically. The majority of electron lenses for TEM use electromagnetic coils to generate a convex lens. The field produced for the lens must be radially symmetrical, as deviation from the radial symmetry of the magnetic lens causes aberrations such as astigmatism, and worsens spherical and chromatic aberration. Electron lenses are manufactured from iron, iron-cobalt or nickel cobalt alloys, such as permalloy. These are selected for their magnetic properties, such as magnetic saturation, hysteresis and permeability.

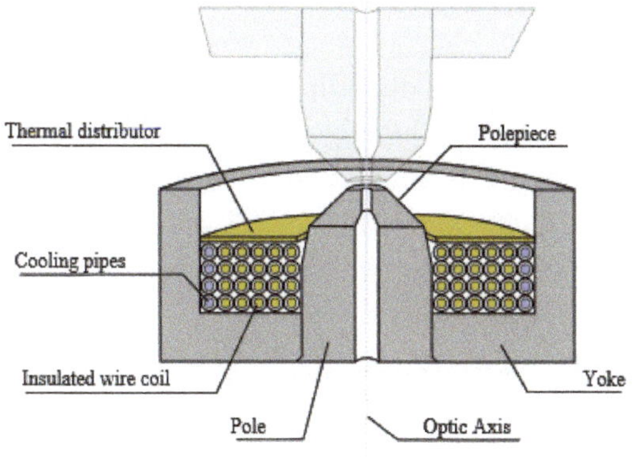

Diagram of a TEM split pole piece design lens

The components include the yoke, the magnetic coil, the poles, the polepiece, and the external control circuitry. The pole piece must be manufactured in a very symmetrical manner, as this provides the boundary conditions for the magnetic field that forms the lens. Imperfections in the manufacture of the pole piece can induce severe distortions in the magnetic field symmetry, which induce distortions that will ultimately limit the lenses' ability to reproduce the object plane. The exact dimensions of the gap, pole piece internal diameter and taper, as well as the overall design of the lens is often performed by finite element analysis of the magnetic field, whilst considering the thermal and electrical constraints of the design.

The coils which produce the magnetic field are located within the lens yoke. The coils can contain a variable current, but typically utilize high voltages, and therefore require significant insulation in order to prevent short-circuiting the lens components. Thermal distributors are placed to ensure the extraction of the heat generated by the energy lost to resistance of the coil windings. The windings may be water-cooled, using a chilled water supply in order to facilitate the removal of the high thermal duty.

Apertures

Apertures are annular metallic plates, through which electrons that are further than a fixed distance from the optic axis may be excluded. These consist of a small metallic disc that is sufficiently thick to prevent electrons from passing through the disc, whilst permitting axial electrons. This permission of central electrons in a TEM causes two effects simultaneously: firstly, apertures decrease the beam intensity as electrons are filtered from the beam, which may be desired in the case of beam sensitive samples. Secondly, this filtering removes electrons that are scattered to high angles, which may be due to unwanted processes such as spherical or chromatic aberration, or due to diffraction from interaction within the sample.

Apertures are either a fixed aperture within the column, such as at the condenser lens, or are a movable aperture, which can be inserted or withdrawn from the beam path, or moved in the plane perpendicular to the beam path. Aperture assemblies are mechanical devices which allow for the selection of different aperture sizes, which may be used by the operator to trade off intensity and the filtering effect of the aperture. Aperture assemblies are often equipped with micrometers to move the aperture, required during optical calibration.

Imaging Methods

Imaging methods in TEM utilize the information contained in the electron waves exiting from the sample to form an image. The projector lenses allow for the correct positioning of this electron wave distribution onto the viewing system. The observed intensity, I, of the image, assuming sufficiently high quality of imaging device, can be approximated as proportional to the time-averaged amplitude of the electron wavefunctions, where the wave that forms the exit beam is denoted by Ψ.

$$I(x) = \frac{k}{t_1 - t_0} \int_{t_0}^{t_1} \Psi \Psi^* dt$$

Different imaging methods therefore attempt to modify the electron waves exiting the sample in a way that provides information about the sample, or the beam itself. From the previous equation, it can be deduced that the observed image depends not only on the amplitude of beam, but also on the phase of the electrons, although phase effects may often be ignored at lower magnifications. Higher resolution imaging requires thinner samples and higher energies of incident electrons, which means that the sample can no longer be considered to be absorbing electrons (i.e., via a Beer's law effect). Instead, the sample can be modeled as an object that does not change the amplitude of the incoming electron wave function, but instead modifies the phase of the incoming wave; in this model, the sample is known as a pure phase object. For sufficiently thin specimens, phase effects dominate the image, complicating analysis of the observed intensities. To improve the contrast in the image, the TEM may be operated at a slight defocus to enhance contrast, owing to convolution by the contrast transfer function of the TEM, which would normally decrease contrast if the sample was not a weak phase object.

Contrast Formation

Contrast formation in the TEM depends greatly on the mode of operation. Complex imaging techniques, which utilize the unique ability to change lens strength or to deactivate a lens, allow for many operating modes. These modes may be used to discern information that is of particular interest to the investigator.

Bright Field

The most common mode of operation for a TEM is the bright field imaging mode. In this mode the contrast formation, when considered classically, is formed directly by occlusion and absorption of electrons in the sample. Thicker regions of the sample, or regions with a higher atomic number will appear dark, whilst regions with no sample in the beam path will appear bright – hence the term "bright field". The image is in effect assumed to be a simple two dimensional projection of the sample down the optic axis, and to a first approximation may be modelled via Beer's law, more complex analyses require the modelling of the sample to include phase information.

Diffraction Contrast

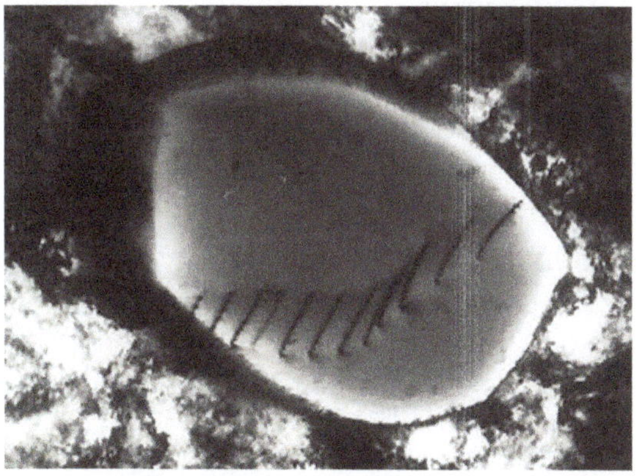

Transmission electron micrograph of dislocations in steel, which are faults in the structure of the crystal lattice at the atomic scale

Samples can exhibit diffraction contrast, whereby the electron beam undergoes Bragg scattering, which in the case of a crystalline sample, disperses electrons into discrete locations in the back focal plane. By the placement of apertures in the back focal plane, i.e. the objective aperture, the desired Bragg reflections can be selected (or excluded), thus only parts of the sample that are causing the electrons to scatter to the selected reflections will end up projected onto the imaging apparatus.

If the reflections that are selected do not include the unscattered beam (which will appear up at the focal point of the lens), then the image will appear dark wherever no sample scattering to the selected peak is present, as such a region without a specimen will appear dark. This is known as a dark-field image.

Modern TEMs are often equipped with specimen holders that allow the user to tilt the specimen to a range of angles in order to obtain specific diffraction conditions, and apertures placed above the specimen allow the user to select electrons that would otherwise be diffracted in a particular direction from entering the specimen.

Applications for this method include the identification of lattice defects in crystals. By carefully selecting the orientation of the sample, it is possible not just to determine the position of defects but also to determine the type of defect present. If the sample is oriented so that one particular plane is only slightly tilted away from the strongest diffracting angle (known as the Bragg Angle), any distortion of the crystal plane that locally tilts the plane to the Bragg angle will produce particularly strong contrast variations. However, defects that produce only displacement of atoms that do not tilt the crystal to the Bragg angle (i. e. displacements parallel to the crystal plane) will not produce strong contrast.

Electron Energy Loss

Utilizing the advanced technique of EELS, for TEMs appropriately equipped, electrons can be rejected based upon their voltage (which, due to constant charge is their energy), using magnetic sector based devices known as EELS spectrometers. These devices allow for the selection of particular energy values, which can be associated with the way the electron has interacted with the sample. For example, different elements in a sample result in different electron energies in the beam after the sample. This normally results in chromatic aberration – however this effect can, for example, be used to generate an image which provides information on elemental composition, based upon the atomic transition during electron-electron interaction.

EELS spectrometers can often be operated in both spectroscopic and imaging modes, allowing for isolation or rejection of elastically scattered beams. As for many images inelastic scattering will include information that may not be of interest to the investigator thus reducing observable signals of interest, EELS imaging can be used to enhance contrast in observed images, including both bright field and diffraction, by rejecting unwanted components.

Phase Contrast

Crystal structure can also be investigated by high-resolution transmission electron microscopy (HRTEM), also known as phase contrast. When utilizing a Field emission source and a specimen of uniform thickness, the images are formed due to differences in phase of electron waves, which is caused by specimen interaction. Image formation is given by the complex modulus of the incoming electron beams. As such, the image is not only dependent on the number of electrons hitting the screen, making direct interpretation of phase contrast images more complex. However this effect can be used to an advantage, as it can be manipulated to provide more information about the sample, such as in complex phase retrieval techniques.

Diffraction

Crystalline diffraction pattern from a twinned grain of FCC Austenitic steel

As previously stated, by adjusting the magnetic lenses such that the back focal plane of the lens rather than the imaging plane is placed on the imaging apparatus a diffraction pattern can be generated. For thin crystalline samples, this produces an image that consists of a pattern of dots in the case of a single crystal, or a series of rings in the case of a polycrystalline or amorphous solid material. For the single crystal case the diffraction pattern is dependent upon the orientation of the specimen and the structure of the sample illuminated by the electron beam. This image provides the investigator with information about the space group symmetries in the crystal and the crystal's orientation to the beam path. This is typically done without utilizing any information but the position at which the diffraction spots appear and the observed image symmetries.

Diffraction patterns can have a large dynamic range, and for crystalline samples, may have intensities greater than those recordable by CCD. As such, TEMs may still be equipped with film cartridges for the purpose of obtaining these images, as the film is a single use detector.

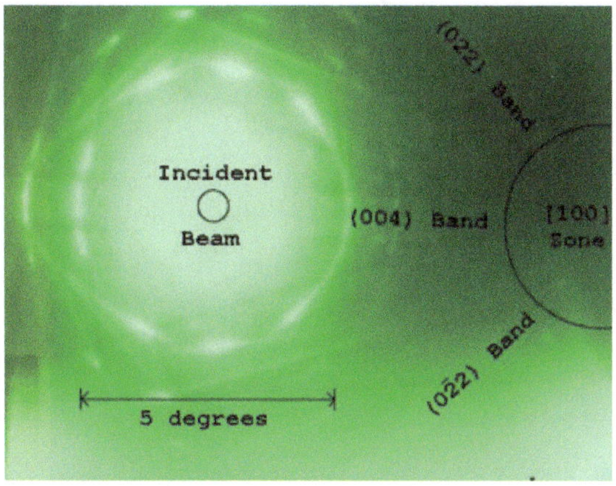

Convergent-beam Kikuchi lines from silicon, near the zone axis

Analysis of diffraction patterns beyond point-position can be complex, as the image is sensitive to a number of factors such as specimen thickness and orientation, objective lens defocus, spherical and chromatic aberration. Although quantitative interpretation of the contrast shown in lattice images is possible, it is inherently complicated and can require extensive computer simulation and analysis, such as electron multislice analysis.

More complex behaviour in the diffraction plane is also possible, with phenomena such as Kikuchi lines arising from multiple diffraction within the crystalline lattice. In convergent beam electron diffraction (CBED) where a non-parallel, i.e. converging, electron wavefront is produced by concentrating the electron beam into a fine probe at the sample surface, the interaction of the convergent beam can provide information beyond structural data such as sample thickness.

Three-dimensional Imaging

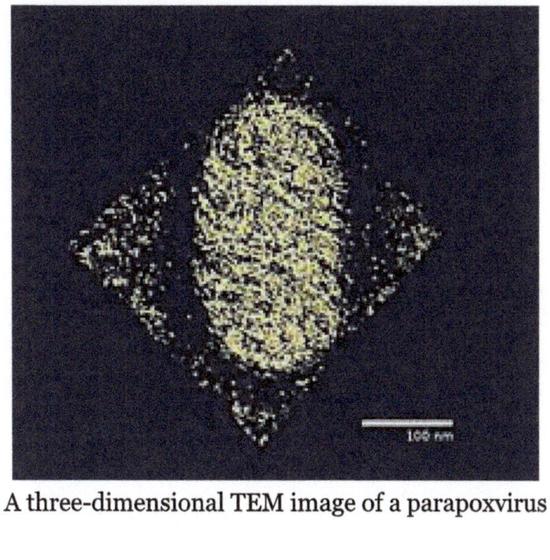

A three-dimensional TEM image of a parapoxvirus

As TEM specimen holders typically allow for the rotation of a sample by a desired angle, multiple views of the same specimen can be obtained by rotating the angle of the sample along an axis perpendicular to the beam. By taking multiple images of a single TEM sample at differing angles, typically in 1° increments, a set of images known as a "tilt series" can be collected. This methodology was proposed in the 1970s by Walter Hoppe. Under purely absorption contrast conditions, this set of images can be used to construct a three-dimensional representation of the sample.

The reconstruction is accomplished by a two-step process, first images are aligned to account for errors in the positioning of a sample; such errors can occur due to vibration or mechanical drift. Alignment methods use image registration algorithms, such as autocorrelation methods to correct these errors. Secondly, using a reconstruction algorithm, such as filtered back projection, the aligned image slices can be transformed from a set of two-dimensional images, $I_j(x, y)$, to a single three-dimensional image, $I'_j(x, y, z)$. This three-dimensional image is of particular interest when morphological information is required, further study can be undertaken using computer algorithms, such as isosurfaces and data slicing to analyse the data.

As TEM samples cannot typically be viewed at a full 180° rotation, the observed images typically suffer from a "missing wedge" of data, which when using Fourier-based back projection methods decreases the range of resolvable frequencies in the three-dimensional reconstruction. Mechanical refinements, such as multi-axis tilting (two tilt series of the same specimen made at orthogonal directions) and conical tomography (where the specimen is first tilted to a given fixed angle and then imaged at equal angular rotational increments through one complete rotation in the plane of the specimen grid) can be used to limit the impact of the missing data on the observed specimen morphology. Using focused ion beam milling, a new technique has been proposed which uses pillar-shaped specimen and a dedicated on-axis tomography holder to perform 180° rotation of the sample inside the pole piece of the objective lens in TEM. Using such arrangements, quantitative electron tomography without the missing wedge is possible. In addition, numerical techniques exist which can improve the collected data.

All the above-mentioned methods involve recording tilt series of a given specimen field. This inevitably results in the summation of a high dose of reactive electrons through the sample and the accompanying destruction of fine detail during recording. The technique of low-dose (minimal-dose) imaging is therefore regularly applied to mitigate this effect. Low-dose imaging is performed by deflecting illumination and imaging regions simultaneously away from the optical axis to image an adjacent region to the area to be recorded (the high-dose region). This area is maintained centered during tilting and refocused before recording. During recording the deflections are removed so that the area of interest is exposed to the electron beam only for the duration required for imaging. An improvement of this technique (for objects resting on a sloping substrate film) is to have two symmetrical off-axis regions for focusing followed by setting focus to the average of the two high-dose focus values before recording the low-dose area of interest.

Non-tomographic variants on this method, referred to as single particle analysis, use images of multiple (hopefully) identical objects at different orientations to produce the image data required for three-dimensional reconstruction. If the objects do not have significant preferred orientations, this method does not suffer from the missing data wedge (or cone) which accompany tomographic methods nor does it incur excessive radiation dosage, however it assumes that the different objects imaged can be treated as if the 3D data generated from them arose from a single stable object.

Sample Preparation

A sample of cells (black) stained with osmium tetroxide and uranyl acetate embedded in epoxy resin (amber) ready for sectioning.

Sample preparation in TEM can be a complex procedure. TEM specimens are required to be at most hundreds of nanometers thick, as unlike neutron or X-Ray radiation the electron beam interacts readily with the sample, an effect that increases roughly with atomic number squared (z^2). High quality samples will have a thickness that is comparable to the mean free path of the electrons that travel through the samples, which may be only a few tens of nanometers. Preparation of TEM specimens is specific to the material under analysis and the desired information to obtain from the specimen. As such, many generic techniques have been used for the preparation of the required thin sections.

Materials that have dimensions small enough to be electron transparent, such as powders or nanotubes, can be quickly prepared by the deposition of a dilute sample containing the specimen onto support grids or films. In the biological sciences in order to withstand the instrument vacuum and facilitate handling, biological specimens can be fixated using either a negative staining material such as uranyl acetate or by plastic embedding. Alternately samples may be held at liquid nitrogen temperatures after embedding in vitreous ice. In material science and metallurgy the specimens tend to be naturally resistant to vacuum, but still must be prepared as a thin foil, or etched so some portion of the specimen is thin enough for the beam to penetrate. Constraints on the thickness of the material may be limited by the scattering cross-section of the atoms from which the material is comprised.

Tissue Sectioning

By passing samples over a glass or diamond edge, small, thin sections can be readily obtained using a semi-automated method. This method is used to obtain thin, minimally deformed samples that allow for the observation of tissue samples. Additionally inorganic samples have been studied, such as aluminium, although this usage is limited owing to the heavy damage induced in the less soft samples. To prevent charge build-up at the sample surface, tissue samples need to be coated with a thin layer of conducting material, such as carbon, where the coating thickness is several nanometers. This may be achieved via an electric arc deposition process using a sputter coating device.

A diamond knife blade used for cutting ultrathin sections (typically 70 to 350 nm) for transmission electron microscopy.

Sample Staining

Details in light microscope samples can be enhanced by stains that absorb light; similarly TEM samples of biological tissues can utilize high atomic number stains to enhance contrast. The stain absorbs electrons or scatters part of the electron beam which otherwise is projected onto the imaging system. Compounds of heavy metals such as osmium, lead, uranium or gold (in immuno-gold labeling) may be used prior to TEM observation to selectively deposit electron dense atoms in or on the sample in desired cellular or protein regions, requiring an understanding of how heavy metals bind to biological tissues.

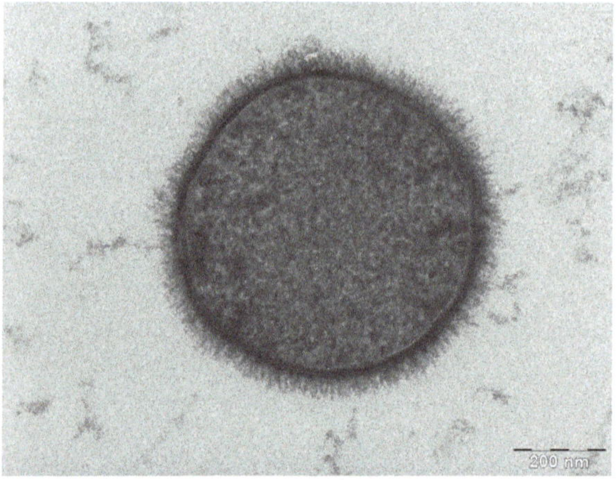

A section of a cell of *Bacillus subtilis*, taken with a Tecnai T-12 TEM. The scale bar is 200 nm.

Mechanical Milling

Mechanical polishing may be used to prepare samples. Polishing needs to be done to a high quality, to ensure constant sample thickness across the region of interest. A diamond, or cubic boron nitride polishing compound may be used in the final stages of polishing to remove any scratches that may cause contrast fluctuations due to varying sample thickness. Even after careful mechanical milling, additional fine methods such as ion etching may be required to perform final stage thinning.

Chemical Etching

Certain samples may be prepared by chemical etching, particularly metallic specimens. These samples are thinned using a chemical etchant, such as an acid, to prepare the sample for TEM observation. Devices to control the thinning process may allow the operator to control either the voltage or current passing through the specimen, and may include systems to detect when the sample has been thinned to a sufficient level of optical transparency.

Ion Etching

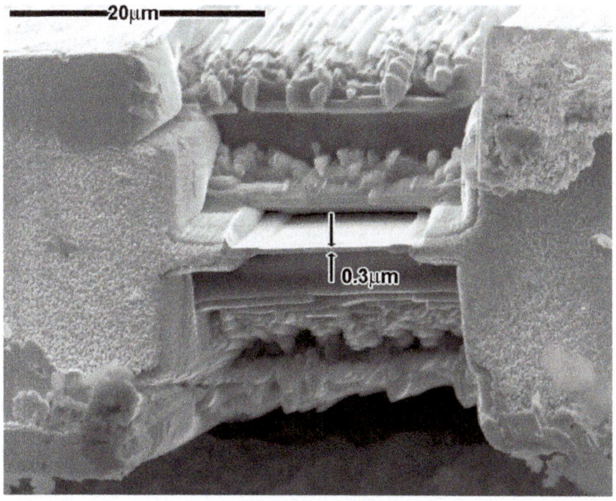

SEM image of a thin TEM sample milled by FIB. The thin membrane shown here is suitable for TEM examination; however, at ~300-nm thickness, it would not be suitable for high-resolution TEM without further milling.

Ion etching is a sputtering process that can remove very fine quantities of material. This is used to perform a finishing polish of specimens polished by other means. Ion etching uses an inert gas passed through an electric field to generate a plasma stream that is directed to the sample surface. Acceleration energies for gases such as argon are typically a few kilovolts. The sample may be rotated to promote even polishing of the sample surface. The sputtering rate of such methods is on the order of tens of micrometers per hour, limiting the method to only extremely fine polishing.

More recently focused ion beam methods have been used to prepare samples. FIB is a relatively new technique to prepare thin samples for TEM examination from larger specimens. Because FIB can be used to micro-machine samples very precisely, it is possible to mill very thin membranes from a specific area of interest in a sample, such as a semiconductor or metal. Unlike inert gas ion sputtering, FIB makes use of significantly more energetic gallium ions and may alter the composition or structure of the material through gallium implantation.

Replication

Samples may also be replicated using cellulose acetate film, the film subsequently coated with a heavy metal, the original film melted away, and the replica imaged on the TEM. This technique is used for both materials and biological samples.

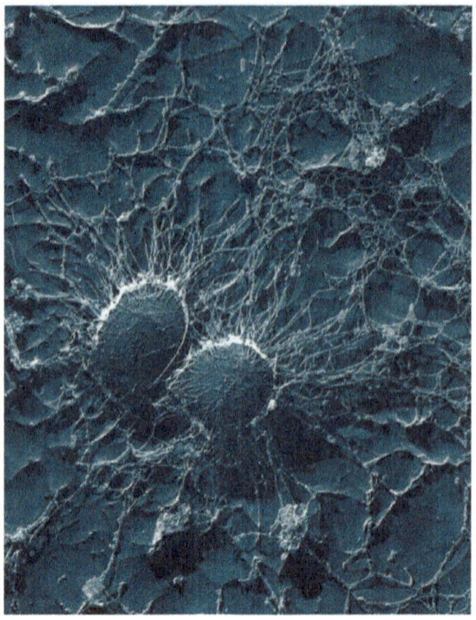

Staphylococcus aureus platinum replica image shot on a TEM at 50,000x magnification

Modifications

The capabilities of the TEM can be further extended by additional stages and detectors, sometimes incorporated on the same microscope.

Scanning TEM

A TEM can be modified into a scanning transmission electron microscope (STEM) by the addition of a system that rasters the beam across the sample to form the image, combined with suitable detectors. Scanning coils are used to deflect the beam, such as by an electrostatic shift of the beam, where the beam is then collected using a current detector such as a Faraday cup, which acts as a direct electron counter. By correlating the electron count to the position of the scanning beam (known as the "probe"), the transmitted component of the beam may be measured. The non-transmitted components may be obtained either by beam tilting or by the use of annular dark field detectors.

Low-voltage Electron Microscope

A low-voltage electron microscope (LVEM) is operated at relatively low electron accelerating voltage between 5–25 kV. Some of these can be a combination of SEM, TEM and STEM in a single compact instrument. Low voltage increases image contrast which is especially important for biological specimens. This increase in contrast significantly reduces, or even eliminates the need to stain. Resolutions of a few nm are possible in TEM, SEM and STEM modes. The low energy of the electron beam means that permanent magnets can be used as lenses and thus a miniature column that does not require cooling can be used.

Cryo-TEM

A cryogenic transmission electron microscope (Cryo-TEM) is a TEM with a specimen holder capable of maintaining the specimen at liquid nitrogen or liquid helium temperatures. This allows imaging specimens prepared in vitreous ice, the preferred preparation technique for imaging individual molecules or macromolecular assemblies, imaging of vitrified solid-electrolye interfaces, and imaging of materials that are volatile in high vacuum at room temperature, such as sulfur.

Environmental/In-situ TEM

In-situ experiments may also be conducted in TEM using differentially pumped sample chambers, or specialized holders. Types of in-situ experiments include studying chemical reactions in liquid cells, and material deformation testing.

Aberration Corrected TEM

Modern research TEMs may include aberration correctors, to reduce the amount of distortion in the image. Incident beam monochromators may also be used which reduce the energy spread of the incident electron beam to less than 0.15 eV. Major aberration corrected TEM manufacturers include JEOL, Hitachi High-technologies, FEI Company, and NION.

Limitations

There are a number of drawbacks to the TEM technique. Many materials require extensive sample preparation to produce a sample thin enough to be electron transparent, which makes TEM analysis a relatively time consuming process with a low throughput of samples. The structure of the sample may also be changed during the preparation process. Also the field of view is relatively small, raising the possibility that the region analyzed may not be characteristic of the whole sample. There is potential that the sample may be damaged by the electron beam, particularly in the case of biological materials.

Resolution Limits

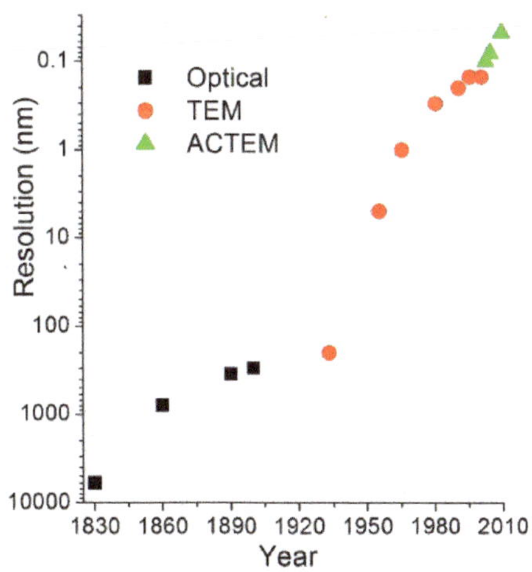

Evolution of spatial resolution achieved with optical, transmission (TEM) and aberration-corrected electron microscopes (ACTEM).

The limit of resolution obtainable in a TEM may be described in several ways, and is typically referred to as the information limit of the microscope. One commonly used value is a cut-off value of the contrast transfer function, a function that is usually quoted in the frequency domain to define the reproduction of spatial frequencies of objects in the object plane by the microscope optics. A cut-off frequency, q_{max}, for the transfer function may be approximated with the following equation, where C_s is the spherical aberration coefficient and λ is the electron wavelength:

$$q_{max} = \frac{1}{0.67(C_s\lambda^3)^{1/4}}.$$

For a 200 kV microscope, with partly corrected spherical aberrations ("to the third order") and a C_s value of 1 μm, a theoretical cut-off value might be $1/q_{max}$ = 42 pm. The same microscope without a corrector would have C_s = 0.5 mm and thus a 200-pm cut-off. The spherical aberrations are suppressed to the third or fifth order in the "aberration-corrected" microscopes. Their resolution is however limited by electron source geometry and brightness and chromatic aberrations in the objective lens system.

The frequency domain representation of the contrast transfer function may often have an oscillatory nature, which can be tuned by adjusting the focal value of the objective lens. This oscillatory nature implies that some spatial frequencies are faithfully imaged by the microscope, whilst others are suppressed. By combining multiple images with different spatial frequencies, the use of techniques such as focal series reconstruction can be used to improve the resolution of the TEM in a limited manner. The contrast transfer function can, to some extent, be experimentally approximated through techniques such as Fourier transforming images of amorphous material, such as amorphous carbon.

More recently, advances in aberration corrector design have been able to reduce spherical aberrations and to achieve resolution below 0.5 Ångströms (50 pm) at magnifications above 50 million times. Improved resolution allows for the imaging of lighter atoms that scatter electrons less efficiently, such as lithium atoms in lithium battery materials. The ability to determine the position of atoms within materials has made the HRTEM an indispensable tool for nanotechnology research and development in many fields, including heterogeneous catalysis and the development of semiconductor devices for electronics and photonics.

Cryo-electron Microscopy

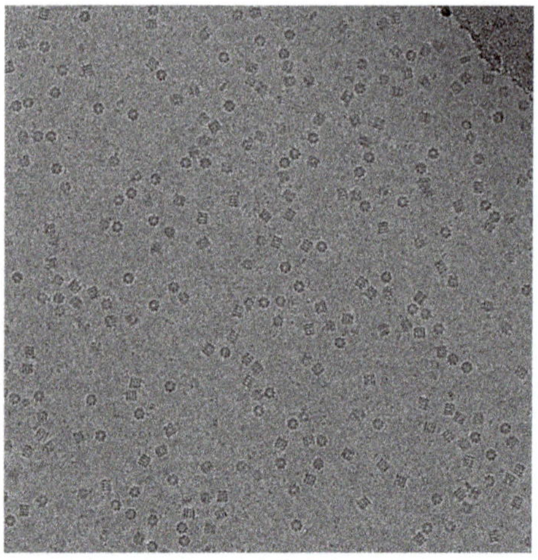

CryoEM image of GroEL suspended in amorphous ice at 50000× magnification

Cryo-electron microscopy (cryo-EM), or electron cryomicroscopy, is a form of transmission electron microscopy (TEM) where the sample is studied at cryogenic temperatures (generally liquid-nitrogen temperatures). Cryo-EM is gaining popularity in structural biology.

The utility of cryoelectron microscopy stems from the fact that it allows the observation of specimens that have not been stained or fixed in any way, showing them in their native environment. This is in contrast to X-ray crystallography, which requires crystallizing the specimen, which can be difficult, and placing them in non-physiological environments, which can occasionally lead to functionally irrelevant conformational changes.

The resolution of cryo-EM maps is improving steadily, and in 2014 some structures at near-atomic resolution had been obtained, including those of viruses, ribosomes, mitochondria, ion channels, and enzyme complexes as small as 170 kD at a resolution of 4.5 Å. Bridget Carragher and colleagues at the Scripps National Resource for Automated Molecular Microscopy used techniques she and Clint Potter developed to create the first cryo-electron microscopy structural biology image with a resolution finer than 3 angstroms, thereby elevating cryo-EM as a tool comparable to and potentially superior to traditional x-ray crystallography techniques. A 2.2 Å map of a bacterial enzyme beta-galactosidase was published in June 2015. A version of electron cryomicroscopy is cryo-electron tomography (CET), where a 3D reconstruction of a sample is created from tilted 2D images.

Development

The original rationale for cryoelectron microscopy was as a means to fight radiation damage for biological specimens. The amount of radiation required to collect an image of a specimen in the electron microscope is high enough to be a potential source of specimen damage for delicate structures. In addition, the high vacuum required on the column of an electron microscope makes the environment for the sample quite harsh.

The problem of the vacuum was partially solved by the introduction of negative stains but even with negative stains biological samples are prone to structural collapse upon dehydration of the specimen. Embedding the samples in ice below the sublimation temperature was a possibility that was contemplated early on, but water tends to arrange into a crystalline lattice of lower density upon freezing and this can destroy the structure of anything that is embedded in it.

In the early '80s, several groups studying solid state physics were attempting to produce vitreous ice by different means, such as high pressure freezing or flash freezing. In a seminal paper in 1984, the group led by Jacques Dubochet at the European Molecular Biology Laboratory showed images of adenovirus embedded in a vitrified layer of water. This paper is generally considered to mark the origin of cryoelectron microscopy, and the technique has been developed to the point of becoming routine at numerous laboratories throughout the world.

The energy of the electrons used for imaging (80-300 kV) is high enough that covalent bonds can be broken. When imaging specimens vulnerable to radiation damage, it is necessary to limit the electron exposure used to acquire the image. These low exposures require that the images of thousands or even millions of identical frozen molecules be selected, aligned, and averaged to obtain high-resolution maps, using specialized software. A significant improvement in structural features was achieved in 2012 by the introduction of direct electron detectors and better computational algorithms.

Biological Specimens

Thin Film

The biological material is spread on an electron microscopy grid and is preserved in a frozen-hydrated state by rapid freezing, usually in liquid ethane near liquid nitrogen temperature. By maintaining specimens at liquid nitrogen temperature or colder, they can be introduced into the high-vacuum of the electron microscope column. Most biological specimens are extremely radiation sensitive, so they must be imaged with low-dose techniques (usefully, the low temperature of cryo-electron microscopy provides an additional protective factor against radiation damage).

Consequently, the images are extremely noisy. For some biological systems it is possible to average images to increase the signal-to-noise ratio and retrieve high-resolution information about the specimen using the technique known as single particle analysis. This approach in general requires that the things being averaged are identical, although some limited conformational heterogeneity can now be studied (e.g. ribosome). Three-dimensional reconstructions from cryo-EM images of protein complexes and viruses have been solved to sub-nanometer or near-atomic resolution, allowing new insights into the structure and biology of these large assemblies.

Analysis of ordered arrays of protein, such as 2-D crystals of transmembrane proteins or helical arrays of proteins, also allows a kind of averaging which can provide high-resolution information about the specimen. This technique is called electron crystallography.

Vitreous Sections

The thin film method is limited to thin specimens (typically < 500 nm) because the electrons cannot cross thicker samples without multiple scattering events. Thicker specimens can be vitrified by plunge freezing (cryofixation) in ethane (up to tens of μm in thickness) or more commonly by high pressure freezing (up to hundreds of μm). They can then be cut in thin sections (40 to 200 nm thick) with a diamond knife in a cryoultramicrotome at temperatures lower than -135 °C (devitrification temperature). The sections are collected on an electron microscope grid and are imaged in the same manner as specimen vitrified in thin film. This technique is called cryo-electron microscopy of vitreous sections (CEMOVIS) or cryo-electron microscopy of frozen-hydrated sections.

Material Specimens

In addition to allowing vitrified biological samples to be imaged, cryo-EM can also be used to image material specimens that are too volatile in vacuum to image using standard, room temperature electron microscopy. For example, vitrified sections of liquid-solid interfaces can be extracted for analysis by cryo-EM, and sulfur, which is prone to sublimation in the vacuum of electron microscopes, can be stabilized and imaged in cryo-EM.

Techniques in Cryoelectron Microscopy

A variety of techniques can be used in cryoelectron microscopy. Popular techniques include:

1. Electron crystallography
    - Analysis of two-dimensional crystals
    - Analysis of helical filaments or tubes
2. Single particle analysis
3. Cryo-electron tomography
4. MicroED

Virus Size

viruses are generally much smaller than the bacteria and its average size varies from 25-300 nm in diameter. They are visible under electron microscope and only the largest and complex viruses are seen under light microscope with high resolution . Among all, the smallest viruses belong to the families *Circoviridae*, *Parvoviridae* and *Picornaviridae* which measure about 20 - 30 nm in diameter while the largest one belongs to *Poxviridae* that measures around 250-300 nm in diameter. Recently, scientists isolated a new form of virus that infects amoeba and grouped it under a separate family *Mimiviridae*. The members of the family *Mimiviridae* range from 400-800 nm in diameter.

On an average a bacterial cell is about 1400 nm in diameter while an average epithelial cell is about 20,000 nm. Considering both viruses and bacteria to be nearly spherical a bacterial cell has a volume about 30,000 times greater than a virus while an epithelial cell is about 60 million times larger.

Size of viruses belonging to different families:

Family	**Size (nm)**
Poxviridae	300
Iridoviridae	135-300
Asfarviridae	170-220
Herpesviridae	150
Adenoviridae	80-100
Polyomaviridae	40-50
Papillomaviridae	55
Hepadnaviridae	50
Circoviridae	12-27
Parvoviridae	15-25
Retroviridae	80-100
Reoviridae	60-80
Birnaviridae	60

Paramyxoviridae	150-250
Rhabdoviridae	100
Filoviridae	80
Bornaviridae	80-100
Orthomyxoviridae	80-120
Bunyaviridae	80-120
Arenaviridae	50-280
Coronaviridae	120-150
Arteriviridae	60-70
Picornaviridae	30
Caliciviridae	30-40
Astroviridae	30
Togaviridae	70
Flaviviridae	40-60

Circoviridae

Circoviridae is a family of viruses. Birds and mammals serve as natural hosts. There are currently 12 species in this family, divided among 2 genera. Diseases associated with this family include: PCV-2: postweaning multisystemic wasting syndrome; CAV: chicken infectious anemia.

Structure

Viruses in Circoviridae are non-enveloped, with icosahedral and round geometries, and T=1 symmetry. The diameter is around 20 nm. Genomes are circular and non-segmented, around 3.8kb in length. The capsid consists of 12 pentagonal trumpet shaped pentamers. There are two main open reading frames arranged in opposite directions that encode the replication (Rep) and capsid (Cap) proteins. Alternative start codons are common in the avian species.

Genus	Structure	Symmetry	Capsid	Genomic arrangement	Genomic segmentation
Gyrovirus	Icosahedral	T=1	Non-enveloped	Circular	Monopartite
Circovirus	Icosahedral	T=1	Non-enveloped	Circular	Monopartite

Life Cycle

Viral replication is nuclear. Entry into the host cell is achieved by penetration into the host cell. Replication follows the ssDNA rolling circle model. DNA templated transcription, with some alternative splicing mechanism is the method of transcription. The virus exits the host cell by nuclear egress, and nuclear pore export. A stem loop structure with a conserved nonanucleotide motif is located at the 5' intergenic region of circovirus genomes and is thought to initiate rolling-cycle replication. Birds and mammals serve as the natural host. Transmission routes are fecal-oral.

Genus	Host	Tissue tropism	Entry	Release	Replication site	Assembly site	Transmission
Gyrovirus	Birds	Chicken: Thymocytes, erythrobalstoid cells; Egg: embryonal tissues and eggshell membrane	Cell receptor endocytosis	Budding	Nucleus	Nucleus	Horizontal: oral-fecal; vertical: bird to egg
Circovirus	Birds; pigs	None	Cell receptor endocytosis	Budding	Nucleus	Nucleus	Horizontal; oral-fecal

Taxonomy

The family Circoviridae contains two genera, *Circovirus* and *Gyrovirus*, and one proposed genus: *Cyclovirus*.

Two yet-uncategorized circovirus-like viruses have been identified in fish—Barbel circovirus (BaCV) 1 and 2. Their genomes are similar in length and contain two major open reading frames similar to the capsid and replication associated protein genes found in other circoviruses.

- Genus *Circovirus*: type species: Porcine circovirus type 1
- Genus *Gyrovirus*: type species: Chicken anaemia virus
- *Proposed* genus *Cyclovirus* : Dragonfly cyclovirus, Florida woods cockroach-associated cyclovirus

Clinical

A cyclovirus — cyclovirus-Vietnam — has been isolated from the cerebrospinal fluid of 25 Vietnamese patients with CNS infections of unknown aetiology. The same virus has been isolated from the faeces of healthy children and also from pigs and chickens. This suggests an orofaecal route of transmission with a possible animal reservoir.

Parvoviridae

The *Parvoviridae* are a family of small, rugged, genetically-compact DNA viruses, known collectively as parvoviruses. There are currently 56 species in the family, divided among 26 genera and two subfamilies. Members of this family infect a wide array of hosts and have been divided into two subfamilies, which infect either vertebrates (the *Parvovirinae*) or invertebrates (*Densovirinae*).

Parvovirus B19 was the first pathogenic human parvovirus to be discovered and is best known for causing a childhood exanthem called "*fifth disease*" (*erythema infectiosum*), although it is also associated with other diseases including arthritis.

Structure

Parvovirus particles (virions) have a durable non-enveloped protein capsid ~20–30 nm in diameter that contains a single copy of the linear single-stranded ~ 5kb DNA genome, which terminates in small imperfect palindromes that fold into dynamic hairpin telomeres. These terminal hairpins are hallmarks of the family, giving rise to the viral origins of DNA replication and mediating multiple steps in the viral life cycle including genome amplification, packaging, and the establishment of transcription complexes. However, they are often refractory to detection by PCR amplification strategies since they tend to induce polymerase strand-switching. Many parvoviruses are exceptionally resistant to inactivation, remaining infectious for months or years after release into the environment.

Viruses in this family have small protein virions that exhibit T=1 icosahedral symmetry. Their capsid shells are assembled from 60 icosahedrally-ordered copies of a single core protein (VP) sequence, but some of these VP proteins also have N-terminal extensions that are not visible in X-ray structures. Biochemical and serological studies indicate that these extensions become successively exposed at the particle surface during virus maturation and cell entry, where they contribute to virion stability and mediate specific steps in cell trafficking. Parvoviruses appear to be unique in encoding a broad spectrum phospholipase A2 (PLA2) activity, typically in the N-terminus of the longest (VP1) subset of their capsid proteins, which is deployed to mediate virion transfer across the lipid bilayer of host cells

Genome

The viral genome is 4–6 kilobases in length and terminates in imperfectly-palindromic hairpin sequences of ~120–500 nucleotides that exhibit genus-specific secondary structures, and can either be identical at the two ends of the genome (homotelomeric) or can differ in size, sequence and predicted secondary structure (heterotelomeric). Homotelomeric viruses package DNA strands of both senses (into separate capsids) whereas heterotelomeric viruses generally package predominantly negative-sense DNA (discussed in references 5 and 6). All parvoviruses encode two major gene complexes: the non-structural (or rep) gene that encodes the replication initiator protein (called NS1 or Rep), and the VP (or cap) gene, which encodes a nested set of ~2–6 size variants derived from the C-terminus of the single VP protein sequence. Members of the *Parvovirinae* also encode a few (1–4) small genus-specific ancillary proteins that are variably distributed throughout the genome, show little sequence homology to each other, and appear to serve an array of different functions in each genus (references 2–6). Viruses in most genera are mono-sense, meaning that both viral genes are transcribed in a single direction from open reading frames in the same (positive-sense) DNA strand, but members of one genus of homotelomeric invertebrate viruses (genus *Ambidensovirus*) show ambisense organization, with the NS and capsid proteins being transcribed in opposite directions from the 5'-ends of the two complementary DNA strands.

The major non-structural protein, NS1, is a site- and strand-specific endonuclease belonging to the HuH protein superfamily, and also carries a AAA+ SF3 helicase domain. NS1 initiates and drives the viral "rolling hairpin" replication mechanism (RHR), which is a linear adaptation of the more-common "rolling-circle" replication strategy used by many small circular prokaryotic and viral replicons. RHR is a unidirectional mechanism that displaces a single, continuous DNA strand, which rapidly folds and refolds to generate a series of concatemeric duplex replication intermediates. Unit length genomes are then excised from these intermediates by the NS1 endonuclease (reviewed in references 5 and 6), and packaged 3'-to-5' into preformed empty capsids driven by the SF3 helicase activity of NS1/Rep.

Life Cycle

Summary

Viral replication is nuclear. Entry into the host cell is achieved by attachment to host receptors, which mediate internalization via endocytosis. Capsids are metastable, undergoing a series of structural shifts during cell entry that sequentially expose peptides carrying PLA2 activity and trafficking signals . These signals ultimately mediate delivery of the intact virion into the cell nucleus, where genome uncoating allows the establishment of viral DNA replication and transcription complexes that rely predominantly upon the synthetic machinery of their host cell. Replication follows the unidirectional strand displacement mechanism discussed above. Packaged virions from viruses in at least two genera (*Protoparvovirus* and *Bocaparvovirus*) have mechanisms that allow mature virions to be trafficked out of viable host cells prior to cell lysis, but members of most other genera are only released into the environment following death and lysis of the infected cell. Natural animal hosts for parvoviruses include a wide range of vertebrates, insects and some crustacea. For viruses in the *Parvovirinae* transmission routes are commonly fecal-oral or respiratory.

Steps of Life Cycle

Attachment and Entry

A virion attaches to receptors on the surface of a potential host cell. In the case of B19 virus the host cell is a red blood cell precursor and the receptor is the blood group P antigen. The virion enters the cell by endocytosis and is released from the endosome into the cytoplasm, where it associates with microtubules and is transported to a nuclear pore. With a diameter of 18–26 nm, the parvovirus virion is small enough to pass through a nuclear pore, unlike the herpesvirus nucleocapsid, though there is evidence that the virion must undergo some structural changes before it can be transported into the nucleus. Nuclear localization signals have been found in the capsid proteins of some parvoviruses.

Single-stranded DNA to Double-stranded DNA

In the nucleus the single-stranded virus genome is converted to dsDNA by a cell DNA polymerase. The ends of the genome are double stranded as a result of base pairing, and at the 3end the –OH group acts as a primer to which the enzyme binds.

Transcription and Translation

The cell RNA polymerase II transcribes the virus genes and cell transcription factors play key roles. The primary transcript(s) undergo various splicing events to produce two size classes of mRNA. The larger mRNAs encode the non-structural proteins and the smaller mRNAs encode the structural proteins. The non-structural proteins are phosphorylated and play roles in the control of gene expression and in DNA replication.

DNA Replication

After and virion assembly conversion of the ssDNA genome to dsDNA, the DNA is replicated by a mechanism called rolling-hairpin replication. This is a leading strand mechanism and sets parvoviruses apart from other DNA viruses, which replicate their genomes through leading and lagging strand synthesis. Procapsids are constructed from the structural proteins and each is filled by a copy of the virus genome, either a (+) DNA or a (–) DNA as appropriate. One of the non-structural proteins functions as a helicase to unwind the dsDNA so that a single strand can enter the procapsid.

Taxonomy

Group: ssDNA

This taxonomy was last updated by the International Committee on Virus Taxonomy (ICTV) *Parvoviridae* Study Group early in 2014, as detailed in references 1 and 12. The family is divided into two subfamilies: *Parvovirinae*, which infect vertebrates and *Densovirinae*, which infect invertebrates. Each subfamily has been subdivided into several genera.

Subfamily Densovirinae:

- Genus *Ambidensovirus*; type species: *Lepidopteran ambidensovirus 1*. Genus includes 6 recognized species that infect insects from multiple orders
- Genus *Brevidensovirus*: type species: *Dipteran brevidensovirus 1*. Genus includes 2 recognized species that infect insects from the order *Diptera*
- Genus *Hepandensovirus*: type species: *Decapod hepandensovirus 1*. Genus includes a single species, infecting decapod shrimp
- Genus *Iteradensovirus*: type species: *Lepidopteran iteradensovirus 1*. Genus includes 5 recognized species that infect insects from the order *Lepidoptera*
- Genus *Penstyldensovirus*: type species: *Decapod penstyldensovirus 1*. Genus includes a single species, infecting decapod shrimp

Subfamily Parvovirinae:

- Genus *Amdoparvovirus*: type species: *Carnivore amdoparvovirus 1*. Genus includes 2 recognized species, infecting mink and fox
- Genus *Aveparvovirus*: type species: *Galliform aveparvovirus 1*. Genus includes a single species, infecting turkeys and chickens
- Genus *Bocaparvovirus*: type species: *Ungulate bocaparvovirus 1*. Genus includes 12 recognized species, infecting mammals from multiple orders, including primates
- Genus *Copiparvovirus*: type species: *Ungulate copiparvovirus 1*. Genus includes 2 recognized species, infecting pigs and cows
- Genus *Dependoparvovirus*: type species: *Adeno-associated dependoparvovirus A*. Genus includes 7 recognized species, infecting mammals, birds or reptiles

- Genus *Erythroparvovirus*: type species: *Primate erythroparvovirus 1*. Genus includes 6 recognized species, infecting mammals, specifically primates, chipmunk or cows
- Genus *Protoparvovirus*: type species: *Rodent protoparvovirus 1*. Genus includes 5 recognized species, infecting mammals from multiple orders, including primates
- Genus *Tetraparvovirus*: type species: *Primate tetraparvovirus 1*. Genus includes 6 recognized species, infecting primates, bats, pigs, cows and sheep

To date, very few viruses from the *Densovirinae* have been studied and sequenced, so the above taxonomy may poorly reflect the true diversity of this subfamily. Currently, all recognized members of genus *Ambidensovirus*, *Brevidensovirus* and *Iterodensovirus* infect insects, while the hepan- and penstyldensoviruses infect decapod shrimp. However, recently two viruses have been isolated that segregate phylogenetically with members of genus *Ambidensovirus* but infect non-insect hosts. One of these, called cherax quadricarinatus densovirus (CqDV), has been shown to infect and kill the freshwater crayfish *Cherax quadricarinatus*, subphylum *Crustacea*, while the other, called sea-star-associated densovirus (SSaDV), has been shown to replicate in and kill sea-stars and perhaps some other members of phylum *Echinodermata*.

Viruses Infecting Humans

Currently, viruses that infect humans are recognized in 5 genera: *Bocaparvovirus* (human bocavirus 1–4, HboV1–4), *Dependoparvovirus* (adeno-associated virus 1–5, AAV1–5), *Erythroparvovirus* (parvovirus B19, B19), *Protoparvovirus* (Bufavirus 1–2, BuV1–2) and *Tetraparvovirus* (human parvovirus 4 G1–3, PARV4 G1–3).

Unclassified Viruses

Since databases contain vast numbers of sequences that might be considered parvoviral in origin but are not real viruses, the *Parvoviridae* Study group cite the following criteria (references 1 and 12) that must be established before a new viral sequence can be considered for recognition in the family: "In order for an agent to be classified in the family *Parvoviridae*, it must be judged to be an authentic parvovirus on the basis of having been isolated and sequenced or, failing this, on the basis of having been sequenced in tissues, secretions, or excretions of unambiguous host origin, supported by evidence of its distribution in multiple individual hosts in a pattern that is compatible with dissemination by infection. The sequence must be in one piece, contain all the nonstructural (NS) and virus particle (VP) coding regions, and meet the size constraints and motif patterns typical of the family."

This means that partial coding sequences or sequences from a single host animal, with no evidence of virus exposure in the rest of the population, will not be considered sufficiently validated. Samples from feces are particularly problematic because they may be derived from food and so do not have an unequivocal host animal. Similarly, samples from aquatic animals that are also present in the surrounding environment are difficult to attribute until they can be shown to directly infect other members of the presumed host species. Despite these caveats, every year many new viruses are identified that will likely merit recognition, so that the published taxonomy always trails studies in the field.

For example, a candidate parvovirus has recently been isolated by sequencing a histocytic sarcoma in a slow loris (*Nycticebus coucang*). The relationship between the virus and the sarcoma was not clear.

Additional Genera

Proposed additional genera include Chapparvovirus and Marinoparvovirus. Confirmation of the genera awaits a decision by the ICTV.

Picornavirus

A picornavirus is a virus belonging to the family *Picornaviridae*, a family of viruses in the order Picornavirales. Vertebrates, including humans, serve as natural hosts. Picornaviruses are nonenveloped viruses that represent a large family of small, cytoplasmic, plus-strand RNA(~7.5kb) viruses with a 30 nm icosahedral capsid. Its genome does not have a lipid membrane. Picornaviruses are found in mammals and birds. There are currently 50 species in this family, divided among 29 genera. These include the Entero-, Aphtho-, Cardio-, Rhino- and Hepatovirus genera. The viruses in this family can cause a range of diseases including paralysis, meningitis, hepatitis and poliomyelitis. Picornaviruses are in Baltimore IV class. Their genome single-stranded (+) sense RNA is that functions as mRNA after entry into the cell and all viral mRNA synthesized is of genome polarity. The mRNA encodes RNA dependent RNA polymerase. This polymerase makes complementary minus strands of RNA, then uses them as templates to make more plus strands. The first synthesized molecule is protein in contrast to all other viruses. So, Overview of picornavirus replication are in order to attachment, entry, translation, transcription/genome replication (one and the same process), assembly and exit.

Diseases

Enteroviruses infect the enteric tract, which is reflected in their name. On the other hand, rhinoviruses infect primarily the nose and the throat. Enteroviruses replicate at 37 °C, whereas rhinoviruses grow better at 33 °C, as this is the lower temperature of the nose. Enteroviruses are stable under acid conditions and thus they are able to survive exposure to gastric acid. In contrast, rhinoviruses are acid-labile (inactivated or destroyed by low pH conditions) and that is the reason why rhinovirus infections are restricted to the nose and throat.

Picornavirus Genera, Species and Serotypes		
Genus	***Species (* signifies type species)***	***Serotypes***
Aphthovirus	*Bovine rhinitis A virus*	2 types: bovine rhinitis A virus (BRAV) 1–2 (formerly bovine rhinovirus 1 & 3)
	Bovine rhinitis B virus	1 type: bovine rhinitis B virus (BRBV) 1 (formerly bovine rhinovirus 2)
	Equine rhinitis A virus	1 type: equine rhinitis A virus (ERAV) 1 (formerly equine rhinovirus 1)
	Foot-and-mouth disease virus *	7 types: O, A, C, Southern African Territories (SAT) 1, SAT 2, SAT 3 and Asia 1
Aquamavirus	*Aquamavirus A*	1 type: seal aquamavirus A1 (SeAV-A1)
Avihepatovirus	*Duck hepatitis A virus*	3 types: duck hepatitis A virus (DHAV) 1–3

Cardiovirus	*Encephalomyocarditis virus* *	2 types: encephalomyocarditis virus (EMCV) 1 & EMCV-2. Note: Columbia SK virus, Maus Elberfeld virus and Mengovirus are strains of EMCV-1.
	Theilovirus	12 types: Theiler's murine encephalomyelitis virus (TMEV), Vilyuisk human encephalomyelitis virus (VHEV), Thera virus (TRV), Saffold virus (SAFV) 1–9
Cosavirus	*Cosavirus A*	24 types: cosavirus A1 (CoSV-A1) to CoSV-A24
Dicipivirus	*Cadicivirus A*	1 type: canine picodicistrovirus 1 (CaPdV-1)
Enterovirus	*Enterovirus A* (formerly Human enterovirus A)	23 types: coxsackievirus A2 (CV-A2), CV-A3, CV-A4, CV-A5, CV-A6, CV-A7, CV-A8, CV-A10, CV-A12, CV-A14, CV-A16, enterovirus (EV) A71, EV-A76, EV-A89, EV-A90, EV-A91, EV-A92, EV-114, EV-A119, SV19, SV43, SV46 & BA13; see also coxsackie A virus
	Enterovirus B (formerly Human enterovirus B)	60 types: coxsackievirus B1 (CV-B1), CV-B2, CV-B3, CV-B4, CV-B5 (incl. swine vesicular disease virus [SVDV]), CV-B6, CV-A9, echovirus 1 (E-1; incl. E-8), E-2, E-3, E-4, E-5, E-6, E-7, E-9 (incl. CV-A23), E-11, E-12, E-13, E-14, E-15, E-16, E-17, E-18, E-19, E-20, E-21, E-24, E-25, E-26, E-27, E-29, E-30, E-31, E-32, E-33, enterovirus B69 (EV-B69), EV-B73, EV-B74, EV-B75, EV-B77, EV-B78, EV-B79, EV-B80, EV-B81, EV-B82, EV-B83, EV-B84, EV-B85, EV-B86, EV-B87, EV-B88, EV-B93, EV-B97, EV-B98, EV-B100, EV-B101, EV-B106, EV-B107, EV-B110 & SA5
	Enterovirus C * (formerly Human enterovirus C)	23 types: poliovirus (PV) 1, PV-2, PV-3, coxsackievirus A1 (CV-A1), CV-A11, CV-A13, CV-A17, CV-A19, CV-A20, CV-A21, CV-A22, CV-A24, EV-C95, EV-C96, EV-C99, EV-C102, EV-C104, EV-C105, EV-C109, EV-C113, EV-C116, EV-C117 & EV-118
	Enterovirus D (formerly Human enterovirus D)	5 types: enterovirus D68 (EV-D68), EV-D70, EV-D94, EV-D111 & EV-D120
	Enterovirus E (formerly Bovine enterovirus group A)	4 types (proposed): enterovirus E1 (EV-E1), EV-E2, EV-E3 & EV-E4
	Enterovirus F (formerly Bovine enterovirus group B)	6 types (proposed): enterovirus F1 (EV-F1), EV-F2, EV-F3, EV-F4, EV-F5 & EV-E6
	Enterovirus G (formerly Porcine enterovirus B)	6 types: enterovirus (EV) G1 to G6 (formerly porcine enterovirus 9–10, 14–16 and ovine enterovirus 1)
	Enterovirus H (formerly Simian enterovirus A)	1 type: enterovirus H1 (EV-H1)
	Enterovirus J	6 types: simian virus 6 (SV6), enterovirus J103 (EV-J103), EV-J108, EV-J112, EV-J115 and EV-J121
	Rhinovirus A (formerly Human rhinovirus A)	77 types: human rhinovirus (HRV) A1, A2, A7, A8, A9, A10, A11, A12, A13, A15, A16, A18, A19, A20, A21, A22, A23, A24, A25, A28, A29, A30, A31, A32, A33, A34, A36, A38, A39, A40, A41, A43, A44, A45, A46, A47, A49, A50, A51, A53, A54, A55, A56, A57, A58, A59, A60, A61, A62, A63, A64, A65, A66, A67, A68, A71, A73, A74, A75, A76, A77, A78, A80, A81, A82, A85, A88, A89, A90, A94, A95, A96, A98, A100, A101, A102 and A103
	Rhinovirus B (formerly Human rhinovirus B)	25 types: human rhinovirus (HRV) B3, B4, B5, B6, B14, B17, B26, B27, B35, B37, B42, B48, B52, B69, B70, B72, B79, B83, B84, B86, B91, B92, B93, B97 and B99
	Rhinovirus C (formerly Human rhinovirus C)	51 types: human rhinovirus (HRV) C1–C51

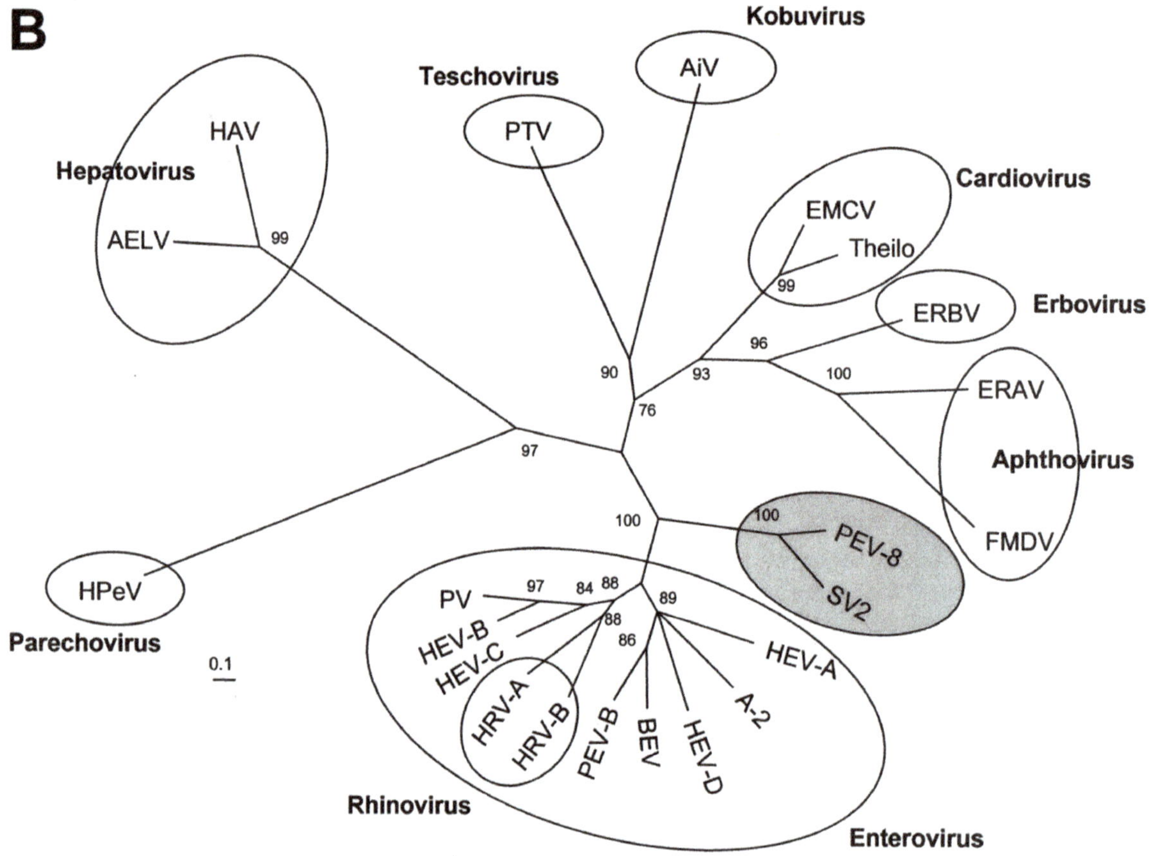

Virology

Structure

Picornaviruses are non-enveloped, with an icosahedral capsid. The capsid is an arrangement of 60 protomers in a tightly packed icosahedral structure. Each protomer consists of 4 polypeptides known as VP (viral protein) 1, 2, 3 and 4. VP2 and VP4 polypeptides originate from one protomer known as VP0 that is cleaved to give the different capsid components. The icosahedral is said to have a triangulation number of 3, this means that in the icosahedral structure each of the 60 triangles that make up the capsid are split into 3 little triangles with a subunit on the corner.In many picornaviruses have a deep cleft formed by around each of the 12 vertices of icosahedrons.The outer surface of the capsid is composed of regions of VP1,VP2 and VP3. Around each of the vertices is a canyon lined with the C termini of VP1 and VP3. The interior surface of the capsid is composed of VP4 and the N termini of VP1. J.Esposito and Professor Freederick A. Murphy demonstrates cleft structure referred to as canyons, using X-ray crystallography and cryo-electron microscopy. Depending on the type and degree of dehydration the viral particle is around 27–30 nm in diameter. The viral genome is around 2500 nm in length so we can therefore conclude that it must be tightly packaged within the capsid along with substances such as sodium ions in order to cancel out the negative charges on the RNA caused by the phosphate groups.

Genus	Structure	Symmetry	Capsid	Genomic Arrangement	Genomic Segmentation
Aphthovirus	Icosahedral	Pseudo T=3	Non-Enveloped	Linear	Monopartite
Aquamavirus	Icosahedral	Pseudo T=3	Non-Enveloped	Linear	Monopartite
Avihepatovirus	Icosahedral	Pseudo T=3	Non-Enveloped	Linear	Monopartite
Avisivirus	Icosahedral	Pseudo T=3	Non-Enveloped	Linear	Monopartite
Cardiovirus	Icosahedral	Pseudo T=3	Non-Enveloped	Linear	Monopartite
Cosavirus	Icosahedral	Pseudo T=3	Non-Enveloped	Linear	Monopartite
Dicipivirus	Icosahedral	Pseudo T=3	Non-Enveloped	Linear	Monopartite
Enterovirus	Icosahedral	Pseudo T=3	Non-Enveloped	Linear	Monopartite
Erbovirus	Icosahedral	Pseudo T=3	Non-Enveloped	Linear	Monopartite
Gallivirus	Icosahedral	Pseudo T=3	Non-Enveloped	Linear	Monopartite
Hepatovirus	Icosahedral	Pseudo T=3	Non-Enveloped	Linear	Monopartite
Hunnivirus	Icosahedral	Pseudo T=3	Non-Enveloped	Linear	Monopartite
Kobuvirus	Icosahedral	Pseudo T=3	Non-Enveloped	Linear	Monopartite
Kunsagivirus	Icosahedral	Pseudo T=3	Non-Enveloped	Linear	Monopartite
Megrivirus	Icosahedral	Pseudo T=3	Non-Enveloped	Linear	Monopartite
Mischivirus	Icosahedral	Pseudo T=3	Non-Enveloped	Linear	Monopartite
Mosavirus	Icosahedral	Pseudo T=3	Non-Enveloped	Linear	Monopartite
Oscivirus	Icosahedral	Pseudo T=3	Non-Enveloped	Linear	Monopartite
Parechovirus	Icosahedral	Pseudo T=3	Non-Enveloped	Linear	Monopartite
Pasivirus	Icosahedral	Pseudo T=3	Non-Enveloped	Linear	Monopartite
Passerivirus	Icosahedral	Pseudo T=3	Non-Enveloped	Linear	Monopartite
Rosavirus	Icosahedral	Pseudo T=3	Non-Enveloped	Linear	Monopartite
Sakobuvirus	Icosahedral	Pseudo T=3	Non-Enveloped	Linear	Monopartite
Salivirus	Icosahedral	Pseudo T=3	Non-Enveloped	Linear	Monopartite
Sapelovirus	Icosahedral	Pseudo T=3	Non-Enveloped	Linear	Monopartite
Senecavirus	Icosahedral	Pseudo T=3	Non-Enveloped	Linear	Monopartite
Sicinivirus	Icosahedral	Pseudo T=3	Non-Enveloped	Linear	Monopartite
Teschovirus	Icosahedral	Pseudo T=3	Non-Enveloped	Linear	Monopartite
Tremovirus	Icosahedral	Pseudo T=3	Non-Enveloped	Linear	Monopartite

Genome

Picornaviruses are classed under Baltimore's viral classification system as group IV viruses as they contain a single stranded, positive sense RNA genome. Their genome ranges between 7.1 and 8.9 kb (kilobases) in length. Like most positive sense RNA genomes, the genetic material alone is infectious; although substantially less virulent than if contained within the viral particle, the RNA can have increased infectivity when transfected into cells. The genome RNA is unusual because it has a protein on the 5' end that is used as a primer for transcription by RNA polymerase.This primer is called VPg genome range between 2–3 kb. VPg contain tyrosine residue at the 3' end. Tyrosine as a –OH source for covalently linked to 5' end of RNA.

The genome is non-segmented and positive-sense (the same sense as mammalian mRNA, being read 5' to 3'). Unlike mammalian mRNA picornaviruses do not have a 5' cap but a virally encoded protein known as VPg. However, like mammalian mRNA, the genome does have a poly(A) tail at the 3' end. There is an un-translated region (UTR) at both ends of the picornavirus genome. The 5' UTR is usually longer, being around 500–1200 nucleotides (nt) in length, compared to that of the 3' UTR, which is around 30–650 nt. It is thought that the 5' UTR is important in translation and the 3' in negative strand synthesis; however the 5' end may also have a role to play in virulence of the virus. The rest of the genome encodes structural proteins at the 5' end and non-structural proteins at the 3' end in a single polyprotein.

The polyprotein is organised as follows: L-1ABCD-2ABC-3ABCD with each letter representing a protein, however, there are variations to this layout.

The 1A, 1B, 1C, and 1D proteins are the capsid proteins VP4, VP2, VP3, and VP1, respectively.Virus-coded proteases perform the cleavages, some of which are intramolecular. The polyprotein is first cut to yield P1, P2 and P3. P1 becomes myristylated at the N terminus before being cleaved to VP0, VP3 and VP1, the proteins that will form procapsids; VP0 will later be cleaved to produce VP2 and VP4. Other cleavage products include 3B (VPg), 2C (an ATPase) and 3D (the RNA polymerase).

Replication

Genomic RNAs of picornaviruses possess multiple RNA elements and they are required for both negative and plus strand RNA synthesis. The cis acting replication(cre) element is a required for replication. The stem-loop-structure contains the cre is independent of position but it changes with location between virus types when it has been identified. Also, the 3' end elements of viral RNA are significant and efficient for RNA replication of picornaviruses. The 3' end of picornavirus contains poly(A) tract which be required for infectivity. On the other hand, RNA synthesis is hypothesized to occur in this region.3' end NCR of poliovirus is not necessary for negative-strands synthesis. However, it is important element for positive—strand synthesis. Additionally,5' end NCR that contain secondary structural elements is required for RNA replication and poliovirus translation initiation(IRES). Internal Ribosome Entry Site (IRES) are RNA structures that allow cap independent initiation of translation, and are able to initiate translation in the middle of a messenger RNA.

The viral particle binds to cell surface receptors. Cell surface receptors are characterized for each serotype of picornaviruses. For example, poliovirus receptor is glycoprotein CD155 which is special receptor for human and some other primate species. For this reason, poliovirus couldn't be made in many laboratories until transgenic mice having a CD155 receptor on their cell surface were developed in the 1990s. These animals can be infected and used for studies of replication and pathogenesis. Binding causes a conformational change in the viral capsid proteins, and myristic acid are released. These acids form a pore in the cell membrane through which RNA is injected . Once inside the cell, the RNA un-coats and the (+) strand RNA genome is replicated through a double-stranded RNA intermediate that is formed using viral RDRP (RNA-Dependent RNA polymerase). Translation by host cell ribosomes is not initiated by a 5' G cap as usual, but rather is initiated by an IRES (Internal Ribosome Entry Site). The viral lifecycle is very rapid with the whole process of replication being completed on average within 8 hours. However, as little as 30 minutes after initial infection, cell protein synthesis declines to almost zero output – essentially the macromolecular synthesis of cell proteins is shut off. Over the next 1–2 hours there is a loss of margination of chromatin and homogeneity in the nucleus, before the viral proteins start to be synthesized and a vacuole appears in the cytoplasm close to the nucleus that gradually starts to spread as the time after infection reaches around 3 hours. After this time the cell plasma membrane becomes permeable, at 4–6 hours the virus particles assemble, and can sometimes be seen in the cytoplasm. At around 8 hours the cell is effectively dead and lyses to release the viral particles.

Experimental data from single step growth-curve-like experiments have allowed scientists to look at the replication of the picornaviruses in great detail. The whole of replication occurs within the host cell cytoplasm and infection can even happen in cells that do not contain a nucleus (known as enucleated cells) and those treated with actinomycin D (this antibiotic would inhibit viral replication if this occurred in the nucleus.)

Translation takes place by -1 ribosomal frameshifting, viral initiation, and ribosomal skipping. The virus exits the host cell by lysis, and viroporins. Vertebrates serve as the natural host. Transmission routes are fecal-oral, contact, ingestion, and air borne particles.

Genus	Host Details	Tissue Tropism	Entry Details	Release Details	Replication Site	Assembly Site	Transmission
Aphthovirus	Ruminants (i.e. cattle, bison, sheep), Horses	Epithelium: soft palate; epithelium: pharynx; epithelium: lung; epithelium: feet; epithelium: mouth	Clathrin-mediated endocytosis	Lysis	Cytoplasm	Cytoplasm	Contact; saliva; aerosol
Aquamavirus	Seals	None	Unknown	Unknown	Cytoplasm	Cytoplasm	Unknown
Avihepatovirus	Birds	None	Unknown	Unknown	Cytoplasm	Cytoplasm	Unknown
Avisivirus	Turkey	None	Unknown	Unknown	Cytoplasm	Cytoplasm	Unknown
Cardiovirus	Humans; vertebrates	Gastrointestinal tract; CNS; heart	Cell receptor endocytosis	Lysis	Cytoplasm	Cytoplasm	Zoonosis; fomite
Cosavirus	Human	None	Cell receptor endocytosis	Unknown	Cytoplasm	Cytoplasm	Unknown
Dicipivirus	Dog	None	Cell receptor endocytosis	Unknown	Cytoplasm	Cytoplasm	Unknown
Enterovirus	Humans; mammals	Gastrointestinal tract	Cell receptor endocytosis	Lysis	Cytoplasm	Cytoplasm	Oral-fecal
Erbovirus	Horse	None	Cell receptor endocytosis	Lysis	Cytoplasm	Cytoplasm	Contact

Genus	Host Details	Tissue Tropism	Entry Details	Release Details	Replication Site	Assembly Site	Transmission
Gallivirus	Turkey, chicken	None	Cell receptor endocytosis	Lysis	Cytoplasm	Cytoplasm	Unknown
Hepatovirus	Humans; vertebrates	Liver	Cell receptor endocytosis	Lysis	Cytoplasm	Cytoplasm	Oral-fecal; blood
Hunnivirus	Cattle	None	Cell receptor endocytosis	Lysis	Cytoplasm	Cytoplasm	Unknown
Kobuvirus	Humans	None	Cell receptor endocytosis	Lysis	Cytoplasm	Cytoplasm	Oral-fecal; blood
Kunsagivirus	Unknown	None	Cell receptor endocytosis	Lysis	Cytoplasm	Cytoplasm	Unknown
Megrivirus	Unknown	None	Cell receptor endocytosis	Lysis	Cytoplasm	Cytoplasm	Unknown
Mischivirus	Unknown	None	Cell receptor endocytosis	Lysis	Cytoplasm	Cytoplasm	Unknown
Mosavirus	Unknown	None	Cell receptor endocytosis	Lysis	Cytoplasm	Cytoplasm	Unknown
Oscivirus	Unknown	None	Cell receptor endocytosis	Lysis	Cytoplasm	Cytoplasm	Unknown
Parechovirus	Humans	Respiratory tract; gastrointestinal tract	Cell receptor endocytosis	Lysis	Cytoplasm	Cytoplasm	Unknown
Pasivirus	Pigs	None	Cell receptor endocytosis	Lysis	Cytoplasm	Cytoplasm	Unknown
Passerivirus	Birds	None	Cell receptor endocytosis	Lysis	Cytoplasm	Cytoplasm	Unknown
Rosavirus	Human, rodents	None	Cell receptor endocytosis	Lysis	Cytoplasm	Cytoplasm	Unknown
Sakobuvirus	Unknown	None	Cell receptor endocytosis	Lysis	Cytoplasm	Cytoplasm	Unknown
Salivirus	Human, chimpanzee	None	Cell receptor endocytosis	Lysis	Cytoplasm	Cytoplasm	Unknown
Sapelovirus	Birds	None	Cell receptor endocytosis	Lysis	Cytoplasm	Cytoplasm	Unknown

Genus	Host Details	Tissue Tropism	Entry Details	Release Details	Replication Site	Assembly Site	Transmission
Senecavirus	Pigs, Cow	None	Cell receptor endocytosis	Lysis	Cytoplasm	Cytoplasm	Unknown
Sicinivirus	Unknown	None	Cell receptor endocytosis	Lysis	Cytoplasm	Cytoplasm	Unknown
Teschovirus	Swine	Gastrointestinal tract; CNS	Cell receptor endocytosis	Lysis	Cytoplasm	Cytoplasm	Oral-fecal
Tremovirus	Birds	None	Unknown	Lysis	Cytoplasm	Cytoplasm	Oral-fecal

VPg

Picornaviruses have a viral protein (VPg) covalently linked to 5' end of their genomes instead of 7-methylguanosine cap like cellular mRNAs. Virus RNA polymerases use VPg as primer. VPg as primer uses both minus and plus strand RNA synthesis. Picornavirus replication is initiated by the uridylylation of viral protein genome-linked (VPg).It is uridylylated at the hydroxyl group of a tyrosine residue. A VPg primer mechanism is utilized by the picornavirus (entero-aphtho- and others), additional virus groups (poty-, como-, calici- and others) and picornavirus-like (coronavirus, notavirus, etc.) supergroup of RNA viruses. The mechanism has been best studied for the enteroviruses (which include many human pathogens, such as poliovirus and coxsackie viruses) as well as for the aphthovirus, an animal pathogen causing foot and mouth disease (FMDV).

In this group, primer-dependent RNA synthesis utilizes a small 22–25 amino acid long viral protein linked to the genome (VPg) to initiate polymerase activity, where the primer is covalently bound to the 5' end of the RNA template. The uridylylation occurs at a tyrosine residue at the third position of the VPg. A cis-acting replication element (CRE), which is a RNA stem loop structure, serves as a template for the uridylylation of VPg, resulting in the synthesis of VPgpUpUOH. Mutations within the CRE-RNA structure prevent VPg uridylylation, and mutations within the VPg sequence can severely diminish RdRp catalytic activity. While the tyrosine hydroxyl of VPg can prime negative-strand RNA synthesis in a CRE- and VPgpUpUOH-independent manner, CRE-dependent VPgpUpUOH synthesis is absolutely required for positive-strand RNA synthesis. CRE-dependent VPg uridylylation lowers the Km¬ of UTP required for viral RNA replication and CRE-dependent VPgpUpUOH synthesis, and is required for efficient negative-strand RNA synthesis, especially when UTP concentrations are limiting. The VPgpUpUOH primer is transferred to the 3' end of the RNA template for elongation, which can continue by addition of nucleotide bases by RdRp. Partial crystal structures for VPgs of foot and mouth disease virus and coxsackie virus B3 suggest that there may be two sites on the viral polymerase for the small VPgs of the picornaviruses. NMR solution structures of poliovirus VPg and VPgpU show that uridylylation stabilizes the structure of the VPg, which is otherwise quite flexible in solution. The second site may be used for uridylylation, after which the VPgpU can initiate RNA synthesis. It should be noted that the VPg primers of caliciviruses, whose structures are only beginning to be revealed, are much larger than those of the picornaviruses. Mechanisms for uridylylation and priming may be quite different in all of these groups.

VPg uridylylation may include the use of precursor proteins, allowing for the determination of a possible mechanism for the location of the diuridylylated, VPg-containing precursor at the 3' end of plus- or minus-strand RNA for production of full-length RNA. Determinants of VPg uridylylation efficiency suggest formation and/or collapse or release of the uridylylated product as the rate-limiting step in vitro depending upon the VPg donor employed. Precursor proteins also have an effect on VPg-CRE specificity and stability. The upper RNA stem loop, to which VPg binds, has a significant impact on both retention, and recruitment, of VPg and Pol. The stem loop of CRE will partially unwind, allowing the precursor components to bind and recruit VPg and Pol4. The CRE loop has a defined consensus sequence to which the initiation components bind, however; there is no consensus sequence for the supporting stem, which suggests that only the structural stability of the CRE is important.

Assembly and organization of the picornavirus VPg ribonucleoprotein complex.

- Step 1: Two 3CD (VPg complex) molecules bind to CRE with the 3C domains (VPg domain) contacting the upper stem and the 3D domains (VPg domain) contacting the lower stem.
- Step 2: The 3C dimer opens the RNA stem by forming a more stable interaction with single strands forming the stem.
- Step 3: 3Dpol is recruited to and retained in this complex by a physical interaction between the back of the thumb subdomain of 3Dpol and a surface of one or both 3C subdomains of 3CD.

VPg may also play an important role in specific recognition of viral genome by movement protein (MP). Movement proteins are non-structural proteins encoded by many, if not all, plant viruses to enable their movement from one infected cell to neighboring cells. MP and VPg interact to provide specificity for the transport of viral RNA from cell to cell. To fulfill energy requirements, MP also interacts with P10, which is a cellular ATPase.

History

In 1897, foot-and-mouth disease virus (FMDV), the first animal virus, was discovered. FMDV is the prototypic member of the *Aphthovirus* genus in the *Picornaviridae* family. The plaque assay was developed using poliovirus; the discovery of viral replication in culture was also with poliovirus in 1949. This was the first time that infection virus had been produced from molecular building blocks in the cells. Polyprotein synthesis, internal ribosome entry sites, and uncapped mRNA were all discovered by studying poliovirus infected cells, and a poliovirus clone was the first infectious DNA clone made of an RNA virus in animals. Along with rhinovirus, poliovirus was the first animal virus to have its structure determined by x-ray crystallography. RNA dependent RNA polymerase was discovered in *Mengovirus*, a genus of picornaviruses.

Related Families

Within the order Picornavirales, there are related viral families, such as the plant infecting Secoviridae, and the insect infecting Dicistroviridae.

Secoviridae

The plant-infecting picorna-like viruses have a number of properties that are distinct from the animal viruses.The picornaviruses include the plant-infecting Secovirida. It has both icosahedral virus particles, viral RNA-dependent RNA polymerase and protease and viral replication proteins. But they have distinguished properties at the same time. For example, secoviruses infect plants and it has specialized proteins. The effect of secoviruses is important on cultivated crops. It infected a wide range of plants from grapevine to rice. They have been classified into the family Secoviridae containing the subfamily Comovirinae (genera Comovirus, Fabavirus and Nepovirus), and genera Sequivirus, Waikavirus, Cheravirus, Sadwavirus, and Torradovirus (type species Tomato torrado virus).

Dicistroviridae

Members of the family dicistroviruses are to beneficial invertebrates such as honey bees and shrimp and to insect pests of medical and agricultural importance that share features with animal and human viruses of the family. Picornaviridae and other insect or marine viruses of the order Picornavirales. Its genome is linear, single stranded positive sense RNA with a viral genome-linked protein (VPg) covalently linked at the 5' end and a 3' poly (A) tract as picornaviruses. Plautia stali intestine virus kelp fly virus, Ectropis obliqua picorna-like virus, deformed wing virus, acute bee paralysis virus, Drosophila C virus, Rhopalosiphum padi virus, and Himetobi P virus. Several have been placed in a separate family—the *Dicistroviridae*. Others have been placed into a new family Iflaviridae. This family includes Infectious flacherie virus and SeIV-1 virus. Another virus is Nora virus from *Drosophila melanogaster*. This latter virus awaits further classification.

Family *Picornaviridae* contains viruses that infect many species of animal as well as humans. Poliovirus was the first virus that was propagated in the cell culture and purified using plaque assay. Most of the picornaviruses grow in a variety of cell lines making them a useful tool to understand the biology of positive strand RNA viruses.

Important Picornaviruses

Poliovirus

In general poliovirus cause mild disease condition in the oro-phrayngeal cavity and gut epithelium. The disease becomes serious when virus migrates to central nervous system and cause systemic viremia. Poliovirus is a disease associated with poor hygienic and sanitary conditions. Poliovirus infection to central nervous system can cause encephalitis and paralysis of limbs and respiratory muscles. Polio vaccine is very effective in controlling the outcome of the disease. Polio has been eradicated from most of the countries.

Hepatitis A virus

Hepatitis A virus is a problem of young children in the developing countries where hygienic condition is not good. Infection in the childrens is often mild and once infected the immunity lasts for life long. The infection to adults can cause severe jaundice which may prove sometimes fatal also.

Coxsackievirus

The first coxsackievirus was isolated from mice showing symptoms of inflammation in skeleton muscles. The virus can cause inflammation of central nervous system and heart muscles with rashes over the body.

Foot and Mouth Disease Virus

Foot and mouth disease (FMD) virus is a highly contagious disease of cloven footed animals. The disease has a major impact on the trade and economy of the countries dependent on agriculture and animal products. The disease causes severe drop in milk production of cattle. Symptoms of the disease include lesions in hoof and mucosal surface of oral cavity. The condition is febrile and animal may die if not treated. Vaccines are available against FMD for cattle, sheep, goats and pigs.

Rhinovirus

Rhinovirus is very common cause of upper respiratory tract infection in humans. Most of the children get infected with rhinovirus by the age of 3 years. Most of the rhinovirus can survive and replicate in the lower temperature (33°C) of respiratory epithelium.

Picornavirus Virion

Picornaviruses are small RNA viruses having diameter of 25-30 nm. The virus has icosahedral symmetry and capsids are made up of four proteins (VP1-4). Some of the picornaviruses contain CANYONS over the vertices of the icosahedron which act as virus attachment site on those viruses (Poliovirus and rhinovirus). Many serotypes have been evolved because of the variation in the capsid proteins which creates a challenge to make a vaccine against the picornaviruses. The genome is composed of ssRNA of 7-8 kb in size. The 5' end of the RNA is covalently attached with a protein called as VPg (genome linked viral protein) while its 3' end is polyadenylated. Interestingly, 5' end of the genome contains many secondary structures called as internal ribosome entry site (IRES).

Picornavirus Replication

Some picornaviruses enter cell through a receptor called as CD155 (Poliovirus receptor). CD155 is a member of immunoglobulin super-family and are widely expressed in many cell types. Many Enteroviruses enter the cell after binding to CD55 or decay accelerating factor, a member of complement system. Once inside the cell the viral encoded VPg gets detached from the 5' end of the RNA. The viral RNA acts as an mRNA and binds to the ribosome with the help of IRES present at the 5' end of the genome. Virus encodes a single polyprotein which is cleaved by virus coded proteases into single structural and nonstructural proteins. The polyprotein is first cleaved into P1, P2, and P3. P1 get cleaved into VP0, VP1 and VP3 and myristylated at its N terminus. VP0 further cleaved into VP2 and VP4, other cleavage products include 2C (ATPase), 3B (VPg) and 3D (RNA polymerase). Virus replication takes place in the replication complexes with the help of RNA polymerase. Usually five copies each of VP0, VP3 and VP1 assembles to form a procapsid which encapsidates the viral genome. Lysis of the infected cells releases the progeny virions.

Schematic representation of picornaviruses genome:

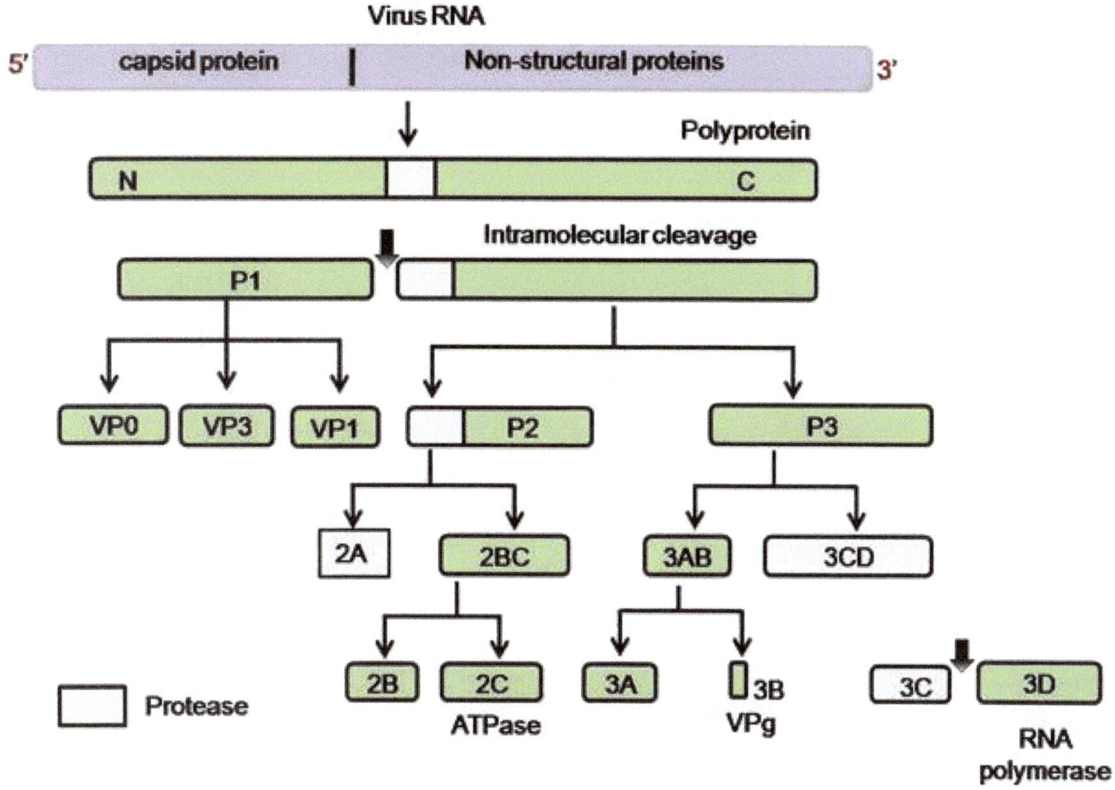

Recombination in Picornavirus

The process of recombination is mostly relevant in cases of segmented viral genome. When a cell is infected simultaneously with two or more strains, the progeny virion may contain the genome derived from both the strains. A recombinant virus is one which contains part of its genome from one virus and remaining from the other. Polio virus vaccine contains the mixture of all three strains responsible for the disease in oral formulation.

References

- Lawrence CM, Menon S, Eilers BJ, et al.. Structural and functional studies of archaeal viruses. The Journal of Biological Chemistry. 2009;284(19):12599–603. doi:10.1074/jbc.R800078200. PMID 19158076
- Casjens S. In: Mahy BWJ and Van Regenmortel MHV. Desk Encyclopedia of General Virology. Boston: Academic Press; 2010. ISBN 0-12-375146-2. p. 167
- Kolata, Gina (2005-10-06). "Experts Unlock Clues to Spread of 1918 Flu Virus". The New York Times. ISSN 0362-4331. Retrieved 2008-02-03
- Rybicki, EP. The classification of organisms at the edge of life, or problems with virus systematics. South African Journal of Science. 1990;86:182–186
- Fenner F.. In: Mahy B. W. J. and Van Regenmortal M. H. V.. Desk Encyclopedia of General Virology. 1 ed. Oxford, UK: Academic Press; 2009. ISBN 0-12-375146-2. p. 15

- Baggesen DL, Sørensen G, Nielsen EM, Wegener HC. Phage typing of Salmonella Typhimurium – is it still a useful tool for surveillance and outbreak investigation?. Eurosurveillance. 2010 [Retrieved 2014-12-19];15(4):19471. PMID 20122382
- Houghton M (November 2009). "The long and winding road leading to the identification of the hepatitis C virus". Journal of Hepatology. 51 (5): 939–48. doi:10.1016/j.jhep.2009.08.004. PMID 19781804
- Mahy WJ & Van Regenmortel MHV (eds). Desk Encyclopedia of General Virology. Oxford: Academic Press; 2009. ISBN 0-12-375146-2. p. 28
- Virus Taxonomy 2013. Archived 26 October 2014 at the Wayback Machine. International Committee on Taxonomy of Viruses. Retrieved on 24 December 2011
- Sanjuán R, Nebot MR, Chirico N, Mansky LM, Belshaw R. Viral mutation rates. Journal of Virology. 2010;84(19):9733–48. doi:10.1128/JVI.00694-10. PMID 20660197
- Mahy WJ & Van Regenmortel MHV (eds). Desk Encyclopedia of General Virology. Oxford: Academic Press; 2009. ISBN 0-12-375146-2. p. 26
- Bianchine PJ, Russo TA. The role of epidemic infectious diseases in the discovery of America. Allergy Proceedings. 1992 [Retrieved 2014-12-19];13(5):225–32. doi:10.2500/108854192778817040. PMID 1483570
- Holmes EC. Viral evolution in the genomic age. PLOS Biology. 2007;5(10):e278. doi:10.1371/journal.pbio.0050278. PMID 17914905
- Belay ED and Schonberger LB. Desk Encyclopedia of Human and Medical Virology. Boston: Academic Press; 2009. ISBN 0-12-375147-0. p. 497–504
- Lodish, Harvey; Berk, Arnold; Zipursky, S. Lawrence; Matsudaira, Paul; Baltimore, David; Darnell, James. Viruses:Structure, Function, and Uses Retrieved on 16 September 2008
- Manipulation of individual viruses: friction and mechanical properties. Biophysical Journal. 1997;72(3):1396–1403. doi:10.1016/S0006-3495(97)78786-1. PMID 9138585. Bibcode: 1997BpJ....72.1396F
- King AMQ, Lefkowitz E, Adams MJ, Carstens EB. Virus Taxonomy: Ninth Report of the International Committee on Taxonomy of Viruses. Elsevier; 2011. ISBN 0-12-384684-6. p. 6
- Pressing J, Reanney DC. Divided genomes and intrinsic noise. Journal of Molecular Evolution. 1984;20(2):135–46. doi:10.1007/BF02257374. PMID 6433032
- Molecular Biology of the Cell; Fourth Edition. New York and London: Garland Science; 2002 [Retrieved 2014-12-19]. ISBN 0-8153-3218-1

2

General Virology: Isolation and Purification

Virus comprises of genomes, proteins and lipids. Genomes contain either DNA or RNA, which helps in synthesis of viral proteins and the replication process. Since viruses require a host body to replicate, they are isolated using methods of cell culture, embryonated eggs and small laboratory animals. Out of these methods, cell culture is mostly preferred for the process. This section has been carefully written to provide an easy understanding of the varied facets of general virology.

Genome

In modern molecular biology and genetics, a genome is the genetic material of an organism. It consists of DNA (or RNA in RNA viruses). The genome includes both the genes (the coding regions), the noncoding DNA and the genetic material of the mitochondria and chloroplasts.

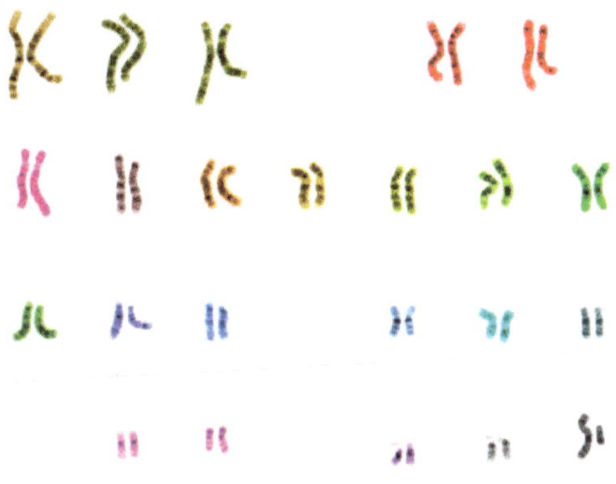

An image of the 46 chromosomes making up the diploid genome of a human male.
(The mitochondrial chromosome is not shown.)

Origin of Term

The term *genome* was created in 1920 by Hans Winkler, professor of botany at the University of Hamburg, Germany. The Oxford Dictionary suggests the name is a blend of the words *gene* and *chromosome*. A few related *-ome* words already existed—such as *biome*, *rhizome*, forming a vocabulary into which *genome* fits systematically.

Overview

Some organisms have multiple copies of chromosomes: diploid, triploid, tetraploid and so on. In classical genetics, in a sexually reproducing organism (typically eukarya) the gamete has half the number of chromosomes of the somatic cell and the genome is a full set of chromosomes in a diploid cell. The halving of the genetic material in gametes is accomplished by the segregation of homologous chromosomes during meiosis. In haploid organisms, including cells of bacteria, archaea, and in organelles including mitochondria and chloroplasts, or viruses, that similarly contain genes, the single or set of circular or linear chains of DNA (or RNA for some viruses), likewise constitute the genome. The term *genome* can be applied specifically to mean what is stored on a complete set of nuclear DNA (i.e., the "nuclear genome") but can also be applied to what is stored within organelles that contain their own DNA, as with the "mitochondrial genome" or the "chloroplast genome". Additionally, the genome can comprise non-chromosomal genetic elements such as viruses, plasmids, and transposable elements.

Typically, when it is said that the genome of a sexually reproducing species has been "sequenced", it refers to a determination of the sequences of one set of autosomes and one of each type of sex chromosome, which together represent both of the possible sexes. Even in species that exist in only one sex, what is described as a "genome sequence" may be a composite read from the chromosomes of various individuals. Colloquially, the phrase "genetic makeup" is sometimes used to signify the genome of a particular individual or organism. The study of the global properties of genomes of related organisms is usually referred to as genomics, which distinguishes it from genetics which generally studies the properties of single genes or groups of genes.

Both the number of base pairs and the number of genes vary widely from one species to another, and there is only a rough correlation between the two (an observation is known as the C-value paradox). At present, the highest known number of genes is around 60,000, for the protozoan causing trichomoniasis, almost three times as many as in the human genome.

An analogy to the human genome stored on DNA is that of instructions stored in a book:

- The book (genome) would contain 23 chapters (chromosomes);
- Each chapter contains 48 to 250 million letters (A,C,G,T) without spaces;
- Hence, the book contains over 3.2 billion letters total;
- The book fits into a cell nucleus the size of a pinpoint;
- At least one copy of the book (all 23 chapters) is contained in most cells of our body. The only exception in humans is found in mature red blood cells which become enucleated during development and therefore lack a genome.

Sequencing and Mapping

In 1976, Walter Fiers at the University of Ghent (Belgium) was the first to establish the complete nucleotide sequence of a viral RNA-genome (Bacteriophage MS2). The next year Fred Sanger completed the first DNA-genome sequence: Phage Φ-X174, of 5386 base pairs. The first complete genome sequences among all three domains of life were released within a short period during the mid-1990s: The first bacterial genome to be sequenced was that of Haemophilus influenzae, completed by a team at The Institute for Genomic Research in 1995. A few months later, the first eukaryotic genome was completed, with sequences of the 16 chromosomes of budding yeast *Saccharomyces cerevisiae* published as the result of a European-led effort begun in the mid-1980s.

The first genome sequence for an archaeon, *Methanococcus jannaschii*, was completed in 1996, again by The Institute for Genomic Research.

```
CATGACGTCGCGGACAACCCAGAATTGTCTTGAGCGATGGTAAGATCTAACCTCACTGCCGGGGGAGGCTCATAC
CTGGGGCTTTACTGATGTCATACCGTCTTGCACGGGGATAGAATGACGGTGCCCGTGTCTGCTTGCCTCGAAGCA
ATTTTCTGAAAGTTACAGACTTCGATTAAAAAGATCGGACTGCGCGTGGGCCCGGAGAGACATGCGTGGTAGTCA
TTTTTCGACGTGTCAAGGACTCAAGGGAATAGTTTGGCGGGAGCGTTACAGCTTCAATTCCCAAAGGTCGCAAGA
CGATAAAATTCAACTACTGGTTTCGGCCTAATAGGTCACGTTTTATGTGAAATAGAGGGGAACCGGCTCCCAAAT
CCCTGGGTGTTCTATGATAAGTCCTGCTTTATAACACGGGGCGGTTAGGTTAAATGACTCTTCTATCTTATGGTG
ATCCAAGCGCCCGCTAATTCTGTTCTGTTAATGTTCATACCAATACTCACATCACATTAGATCAAAGGATCCCCG
AGCCCAGTCGCAAGGGTCTGCTGCTGTTGTCGACGCCTCATGTTACTCCTGGAATCTACCTGCCCTCCCCTCACC
GGTTAAGGCGTGTGATCGACGATGCAGGTATACATCGGCTCGGACCTACAGTGGTCGATCGACTGGCTACTGGCT
TCGCGGTTCGGCGCGTAGTTGAGTGCGATAACCCAACCGGTGGCAAGTAGCAAGAAGACCTACCTGGGTCACCTT
AGACAACCTAACTAATAGTCTCTAACGGGGAATTACCTTTACCAGTCTCATGCCTCCAATATATCTGCACCGCTT
CAATGATATCGCCCACAGAAAGTAGGGTCTCAGGTATCGCATACGCCGCGCCCGGGTCCCAGCTACGCTCAGGAC
GACAGTAGAGAGCTATTGTGTAATTCAGGCTCAGCATTCATCGACCTTTCCTGTTGTGAATATTGTGCTAATGCA
TCTCGTCCGTAACGATCTGGGGGGCAAAACCGAATATCCGTATTCTCGTCCTACGGGTCCACAATGAGAAAGTCC
TGCGCGTGATCGTCAGTTAAGTTAAATTAATTCAGGCTACGGTAAACTTGTAGTGAGCTAAGAATCACGGGAATC
ACGGGTTCGCTACAGATGAACTGAATTTATACACGGACAACTCATCGCCCATTTGGGCGTGGGCACCGCAGATCA
AAAGTGGCAGATTAGGAGTGCTTGATCAGGTTAGCAGGTGGACTGTATCCAACAGCGCATCAAACTTCAATAAAT
CCAAAGCGTTGTAGTGGTCTAAGCACCCCTGAACAGTGGCGCCCATCGTTAGCGTAGTACAACCCTTCCCCCTTG
AGGTGCGACATGGGGCCAGTTAGCCTGCCCTATATCCCTTGCACACGTTCAATAAGAGGGGCTCTACAGCGCCGC
TTTTTAAATTAGGATGCCGACCCCATCATTGGTAACTGTATGTTCATAGATATTTCTTCAGGAGTAATAGCGACA
AGCTGACACGCAAGGGTCAACAATAATTTCTACTATCACCCCGCTGAACGACTGTCTTTGCAAGAACCAACTGGG
CTTAGATTCGCGTCCTAACGTAGTGAGGGCCGAGTCATATCATAGATCAGGCATGAGAAACCGACGTCGAGTCTA
CACACGAGTTGTAAACAACTTGATTGCTATACTGTAGCTACCGCAAGGATCTCCTACATCAAAGACTACTGGGCG
ATCTGGATCCGAGTCAGAAATACGAGTTAATGCAAATTTACGTAGACCGGTGAAAACACGTGCCATGGGTTGCGT
AGACCGTAGTCAGAAGTGTGGCGCGCTATTCGTACCGAACCGGTGGAGTATACAGAATTGCTCTTCTACGACGTA
AGGAGCTCGGTCCCCAATGCACGCCAAAAAAGGAATAAAGTATTCAAACTGCGCATGGTCCCTCCGCCGGTGGCA
CTATTATCCATCCGAACGTTGAACCTACTTCCTCGGCTTATGCTGTCCTCAACAGTATCGCTTATGAATCGCATG
CGGCTGTGGATCTTAACGGCCACATTCTTAATTCCGACCGATCACCGATCGCCTTTCCTCGCTGGTACAATGAGT
ACTAAGTTATCCAGATCAAGGTTTGAACGGACTCGTATGACATGTGTGACTGAACCCGGGAGGAAATGCAGAGAA
CTGTTTCAAGGCCTCTGCTTTGGTATCACTCAATATATTCAGACCAGACAAGTGGCAAAATTTCGTGCGCCTCTC
CTAGGTATTCACGCAACCGTCGTAACATGCACTAAGGATAACTAGCGCCAGGGGGGCATACTAGGTCCCGGAGCT
AAAGACTACCCTATGGATTCCTTGGAGCGGGGACAATGCAGACCGGTTACGACACAATTATCGGGATCGTCTAGA
GGTATTATTAGCAAGACAATAAAGGACATTGCACAGAGACTTATTAGAATTCAACAAACAGGATCATATCATGCG
GTGTTGGGTCGGGCAAGTCCCCGAAGCTCGGCCAAAAGATTCGCCATGGAACCGTCTGGTCCTGTTAGCGTGTAC
GCCTGCTCCTGTTCCGGGTACCATAGATAGACTGAGATTGCGTCAAAAAATTGCGGCGAAAATAGAGGGGCTCCT
TGTAGAAATACCAGACTGGGGAATTTAAGCGCTTTCCACTATCTGAGCGACTAAACATCAACAAATGCGTCTACT
CGAATCCGCAGTAGGCAATTACAACCTGGTTCAGATCACTGGTTAATCAGGGATGTCTTCATAAGATTATACTTG
CCCCGACGCGACAGCTCTTCAAGGGGCCGATTTTTGGACTTCAGATACGCTAGAATTTAAAGGGTCTCTTACACC
TGCTGCGGCCTGCAGGGACCCCTAGAACTTGCCGCCTACTTGTCTCAGTCTAATAACGCGCGAAGCCGTGGGGCA
CGTGACCTTAAGTCGCAGAGCGAGTGATGAATTTGGGACGCTAATATGGGTGAATAGAGACTTATATCATCAGGG
```

Part of DNA sequence - prototypification of complete genome of virus

The development of new technologies has made it dramatically easier and cheaper to do sequencing, and the number of complete genome sequences is growing rapidly. The US National Institutes of Health maintains one of several comprehensive databases of genomic information. Among the thousands of completed genome sequencing projects include those for rice, a mouse, the plant *Arabidopsis thaliana*, the puffer fish, and the bacteria E. coli. In December 2013, scientists first sequenced the entire *genome* of a Neanderthal, an extinct species of humans. The genome was extracted from the toe bone of a 130,000-year-old Neanderthal found in a Siberian cave.

New sequencing technologies, such as massive parallel sequencing have also opened up the prospect of personal genome sequencing as a diagnostic tool, as pioneered by Manteia Predictive Medicine. A major step toward that goal was the completion in 2007 of the full genome of James D. Watson, one of the co-discoverers of the structure of DNA.

Whereas a genome sequence lists the order of every DNA base in a genome, a genome map identifies the landmarks. A genome map is less detailed than a genome sequence and aids in navigating around the genome. The Human Genome Project was organized to map and to sequence the human genome. A fundamental step in the project was the release of a detailed genomic map by Jean Weissenbach and his team at the Genoscope in Paris.

Reference genome sequences and maps continue to be updated, removing errors and clarifying regions of high allelic complexity. The decreasing cost of genomic mapping has permitted genealogical sites to offer it as a service, to the extent that one may submit one's genome to crowd sourced scientific endeavours such as DNA.land at the New York Genome Center, an example both of the economies of scale and of citizen science.

Genome Compositions

Genome composition is used to describe the make up of contents of a haploid genome, which should include genome size, proportions of non-repetitive DNA and repetitive DNA in details. By comparing the genome compositions between genomes, scientists can better understand the evolutionary history of a given genome.

When talking about genome composition, one should distinguish between prokaryotes and eukaryotes as there are significant differences with contents structure. In prokaryotes, most of the genome (85–90%) is non-repetitive DNA, which means coding DNA mainly forms it, while non-coding regions only take a small part. On the contrary, eukaryotes have the feature of exon-intron organization of protein coding genes; the variation of repetitive DNA content in eukaryotes is also extremely high. In mammals and plants, the major part of the genome is composed of repetitive DNA.

Most biological entities that are more complex than a virus sometimes or always carry additional genetic material besides that which resides in their chromosomes. In some contexts, such as sequencing the genome of a pathogenic microbe, "genome" is meant to include information stored on this auxiliary material, which is carried in plasmids. In such circumstances then, "genome" describes all of the genes and information on non-coding DNA that have the potential to be present.

In eukaryotes such as plants, protozoa and animals, however, "genome" carries the typical connotation of only information on chromosomal DNA. So although these organisms contain chloroplasts or mitochondria that have their own DNA, the genetic information contained in DNA within these organelles is not considered part of the genome. In fact, mitochondria are sometimes said to have their own genome often referred to as the "mitochondrial genome". The DNA found within the chloroplast may be referred to as the "plastome".

Genome Size

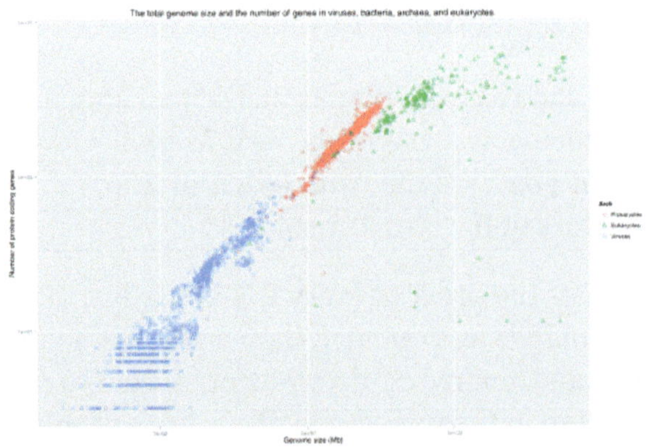

Log-log plot of the total number of annotated proteins in genomes submitted to GenBank as a function of genome size.

Genome size is the total number of DNA base pairs in one copy of a haploid genome. In humans, the nuclear genome comprises approximately 3.2 billion nucleotides of DNA, divided into 24 linear molecules, the shortest 50 000 000 nucleotides in length and the longest 260 000 000 nucleotides, each contained in a different chromosome. The genome size is positively correlated with the morphological complexity among prokaryotes and lower eukaryotes; however, after mollusks and all the other higher eukaryotes above, this correlation is no longer effective. This phenomenon also indicates the mighty influence coming from repetitive DNA act on the genomes.

Since genomes are very complex, one research strategy is to reduce the number of genes in a genome to the bare minimum and still have the organism in question survive. There is experimental work being done on minimal genomes for single cell organisms as well as minimal genomes for multi-cellular organisms. The work is both *in vivo* and *in silico*.

Here is a table of some significant or representative genomes.

Organism type	**Organism**	**Genome size (base pairs)**		**Approx. no. of genes**	**Note**
Virus	Porcine circovirus type 1	1,759	1.8kb		Smallest viruses replicating autonomously in eukaryotic cells.
Virus	Bacteriophage MS2	3,569	3.5kb		First sequenced RNA-genome
Virus	SV40	5,224	5.2kb		
Virus	Phage Φ-X174	5,386	5.4kb		First sequenced DNA-genome
Virus	HIV	9,749	9.7kb		
Virus	Phage λ	48,502	48kb		Often used as a vector for the cloning of recombinant DNA.
Virus	Megavirus	1,259,197	1.3Mb		Until 2013 the largest known viral genome.
Virus	*Pandoravirus salinus*	2,470,000	2.47Mb		Largest known viral genome.
Bacterium	*Nasuia deltocephalinicola* (strain NAS-ALF)	112,091	112kb		Smallest non-viral genome.
Bacterium	*Carsonella ruddii*	159,662	160kb		
Bacterium	*Buchnera aphidicola*	600,000	600kb		
Bacterium	*Wigglesworthia glossinidia*	700,000	700Kb		
Bacterium	*Haemophilus influenzae*	1,830,000	1.8Mb		First genome of a living organism sequenced, July 1995
Bacterium	*Escherichia coli*	4,600,000	4.6Mb	4288	
Bacterium	*Solibacter usitatus* (strain Ellin 6076)	9,970,000	10Mb		
Bacterium – cyanobacterium	*Prochlorococcus* spp. (1.7 Mb)	1,700,000	1.7Mb	1884	Smallest known cyanobacterium genome
Bacterium – cyanobacterium	*Nostoc punctiforme*	9,000,000	9Mb	7432	7432 "open reading frames"

Amoeboid	*Polychaos dubium* (*"Amoeba" dubia*)	670,000,000,000	670Gb		Largest known genome. (Disputed)
Plant	*Genlisea tuberosa*	61,000,000	61Mb		Smallest recorded flowering plant genome, 2014.
Plant	*Arabidopsis thaliana*	135,000,000	135 Mb	27,655	First plant genome sequenced, December 2000.
Plant	*Populus trichocarpa*	480,000,000	480Mb	73013	First tree genome sequenced, September 2006
Plant	*Fritillaria assyrica*	130,000,000,000	130Gb		
Plant	*Paris japonica* (Japanese-native, pale-petal)	150,000,000,000	150Gb		Largest plant genome known
Plant – moss	*Physcomitrella patens*	480,000,000	480Mb		First genome of a bryophyte sequenced, January 2008.
Fungus – yeast	*Saccharomyces cerevisiae*	12,100,000	12.1Mb	6294	First eukaryotic genome sequenced, 1996
Fungus	*Aspergillus nidulans*	30,000,000	30Mb	9541	
Nematode	*Pratylenchus coffeae*	20,000,000	20Mb		Smallest animal genome known
Nematode	*Caenorhabditis elegans*	100,300,000	100Mb	19000	First multicellular animal genome sequenced, December 1998
Insect	*Drosophila melanogaster* (fruit fly)	175,000,000	175Mb	13600	Size variation based on strain (175-180Mb; standard *y w* strain is 175Mb)
Insect	*Apis mellifera* (honey bee)	236,000,000	236Mb	10157	)
Insect	*Bombyx mori* (silk moth)	432,000,000	432Mb	14623	14,623 predicted genes
Insect	*Solenopsis invicta* (fire ant)	480,000,000	480Mb	16569	
Mammal	*Mus musculus*	2,700,000,000	2.7Gb	20210	
Mammal	*Homo sapiens*	3,289,000,000	3.3Gb	20000	*Homo sapiens* estimated genome size 3.2 billion bp Initial sequencing and analysis of the human genome
Mammal	*Bonobo*	3,286,640,000	3.3Gb	20000	*Pan paniscus* estimated genome size 3.29 billion bp
Fish	*Tetraodon nigroviridis* (type of puffer fish)	385,000,000	390Mb		Smallest vertebrate genome known estimated to be 340 Mb – 385 Mb.
Fish	*Protopterus aethiopicus* (marbled lungfish)	130,000,000,000	130Gb		Largest vertebrate genome known

Proportion of non-repetitive DNA

The proportion of non-repetitive DNA is calculated by using the length of non-repetitive DNA divided by genome size. Protein-coding genes and RNA-coding genes are generally non-repetitive

DNA. A bigger genome does not mean more genes, and the proportion of non-repetitive DNA decreases along with increasing genome size in higher eukaryotes.

It had been found that the proportion of non-repetitive DNA can vary a lot between species. Some *E. coli* as prokaryotes only have non-repetitive DNA, lower eukaryotes such as *C. elegans* and fruit fly, still possess more non-repetitive DNA than repetitive DNA. Higher eukaryotes tend to have more repetitive DNA than non-repetitive ones. In some plants and amphibians, the proportion of non-repetitive DNA is no more than 20%, becoming a minority component.

Proportion of Repetitive DNA

The proportion of repetitive DNA is calculated by using length of repetitive DNA divide by genome size. There are two categories of repetitive DNA in genome: tandem repeats and interspersed repeats.

Tandem Repeats

Tandem repeats are usually caused by slippage during replication, unequal crossing-over and gene conversion, satellite DNA and microsatellites are forms of tandem repeats in the genome. Although tandem repeats count for a significant proportion in genome, the largest proportion in mammalian is the other type, interspersed repeats.

Interspersed Repeats

Interspersed repeats mainly come from transposable elements (TEs), but they also include some protein coding gene families and pseudogenes. Transposable elements are able to integrate into the genome at another site within the cell. It is believed that TEs are an important driving force on genome evolution of higher eukaryotes. TEs can be classified into two categories, Class 1 (retrotransposons) and Class 2 (DNA transposons).

Retrotransposons

Retrotransposons can be transcribed into RNA, which are then duplicated at another site into the genome. Retrotransposons can be divided into Long terminal repeats (LTRs) and Non-Long Terminal Repeats (Non-LTR).

Long Terminal Repeats (LTRs)

> Similar to retroviruses, which have both gag and pol genes to make cDNA from RNA and proteins to insert into genome, but LTRs can only act within the cell as they lack the env gene in retroviruses. It has been reported that LTRs consist of the largest fraction in most plant genome and might account for the huge variation in genome size.

Non-long Terminal Repeats (Non-LTRs)

> Can be divided into long interspersed elements (LINEs), short interspersed elements (SINEs) and Penelope-like elements. In *Dictyostelium discoideum*, there is another DIRS-like elements belong to Non-LTRs. Non-LTRs are widely spread in eukaryotic genomes.

Long Interspersed Elements (LINEs)

Are able to encode two Open Reading Frames (ORFs) to generate transcriptase and endonuclease, which are essential in retrotransposition. The human genome has around 500,000 LINEs, taking around 17% of the genome.

Short Interspersed Elements (SINEs)

Are usually less than 500 base pairs and need to co-opt with the LINEs machinery to function as nonautonomous retrotransposons. The Alu element is the most common SINEs found in primates, it has a length of about 350 base pairs and takes about 11% of the human genome with around 1,500,000 copies.

DNA Transposons

DNA transposons generally move by "cut and paste" in the genome, but duplication has also been observed. Class 2 TEs do not use RNA as intermediate and are popular in bacteria, in metazoan it has also been found.

Genome Evolution

Genomes are more than the sum of an organism's genes and have traits that may be measured and studied without reference to the details of any particular genes and their products. Researchers compare traits such as karyotype (chromosome number), genome size, gene order, codon usage bias, and GC-content to determine what mechanisms could have produced the great variety of genomes that exist today.

Duplications play a major role in shaping the genome. Duplication may range from extension of short tandem repeats, to duplication of a cluster of genes, and all the way to duplication of entire chromosomes or even entire genomes. Such duplications are probably fundamental to the creation of genetic novelty.

Horizontal gene transfer is invoked to explain how there is often an extreme similarity between small portions of the genomes of two organisms that are otherwise very distantly related. Horizontal gene transfer seems to be common among many microbes. Also, eukaryotic cells seem to have experienced a transfer of some genetic material from their chloroplast and mitochondrial genomes to their nuclear chromosomes.

Components of Genomes

In general the viruses are made up of nucleic acids (genome), proteins (capsid), and lipids (envelope). Viral genomes can be either DNA or RNA , when once inside a host cell it directs synthesis of new viral proteins, and replication of new viral genomes. Capsid is a protein covering that surrounds and protects the viral genome. It is made up of many small subunits called as capsomeres which determine the shape of the virus. The arrangement and composition of the capsomeres varies among the virus families. Envelopes are the lipid bilayer membranes that are derived from the host cell membrane when virus "buds" out from the plasma membrane or passes through a membrane-bound organelle (such as the Golgi body or endoplasmic reticulum). The envelope contains sometimes glycoprotein (protein with carbohydrate) in the form of spikes which helps them in the attachment during the time of infection to the host cell surface (gp120 in HIV). In non-enveloped viruses, grooves present in the capsid and specific capsid proteins may bind to the cell surface receptor.

The most important and characteristic feature of a living organism is replication of its genetic information. The mechanism of genome replication is done with greater economy and simplicity among different viruses. Different families of viruses have their genome made of either double stranded (ds) DNA or single stranded (ss) DNA or RNA. The viruses that contain RNA genome may have either positive, negative, or mixed (ambisense) polarity. In addition, they either have single or multiple segments in their genome with linear or circular topology. Each of the above parameters have their consequences for the pathways of viral genome replication, viral gene expression, and virion assembly.

Among the families of viruses that infect animals and human, those containing RNA genome outnumber those containing DNA genome. This disparity is even more in case of plant viruses (no double stranded DNA virus that infect plant is known).

Viruses Encode Enzymes and Follow Unique Pathways:

Almost all viruses encodes unique proteins and enzymes, moreover they follow unique pathways to transfer their genetic information. This phenomenon is more pronounced in case of RNA viruses, they either use RNA dependent RNA polymerase or in case of retrovirus (HIV) RNA dependent DNA polymerase to complete their replication cycle. Both of these processes requires unique enzymes that are encoded by the virus following infection to the host cells and are generally absent elsewhere.

The RNA dependent RNA polymerase and reverse transcriptase have minimal proofreading ability, as a result their error rate is very high (1 in 10,000) as compared to the DNA replication. This means that an RNA virus particle will contain 1 or more mutation from its parental wild type virus. Presence of many different subspecies of virus particle in a population is also called as quasispecies nature of RNA viruses. The error prone activity of RNA virus polymerase restricts the upper size limit of the genome above which they cannot survive. As a result of this phenomenon most of the RNA virus have their genome size in the range of 5-15kb (coronavirus 30kb). The opposite is true in case of DNA viruses where proofreading and error repair activity ensures accurate replication of the viral DNA as big as 800 kb. The fact that DNA is more stable chemically than RNA likely explains us why all thermophilic hosts contain viruses that have dsDNA as their genetic material.

Nature of genome of viruses belonging to different families:

Family	**Nature of Genome**
Poxviridae	dsDNA
Iridoviridae	dsDNA
Asfarviridae	dsDNA
Herpesviridae	dsDNA
Adenoviridae	dsDNA
Polyomaviridae	dsDNA

Papillomaviridae	dsDNA
Hepadnaviridae	dsDNA-RT
Circoviridae	ssDNA
Parvoviridae	ssDNA
Retroviridae	ssRNA-RT
Reoviridae	dsRNA
Birnaviridae	dsRNA
Paramyxoviridae	NssRNA
Rhabdoviridae	NssRNA
Filoviridae	NssRNA
Bornaviridae	NssRNA
Orthomyxoviridae	NssRNA
Bunyaviridae	NssRNA
Arenaviridae	NssRNA
Coronaviridae	ssRNA
Arteriviridae	ssRNA
Picornaviridae	ssRNA
Caliciviridae	ssRNA
Astroviridae	ssRNA
Togaviridae	ssRNA
Flaviviridae	ssRNA

dsDNA= double stranded DNA

ssDNA= single stranded DNA

dsRNA= double stranded RNA

ssRNA= single stranded RNA

NssRNA= single stranded RNA with negative polarity

Diversity among the viruses belonging to different groups

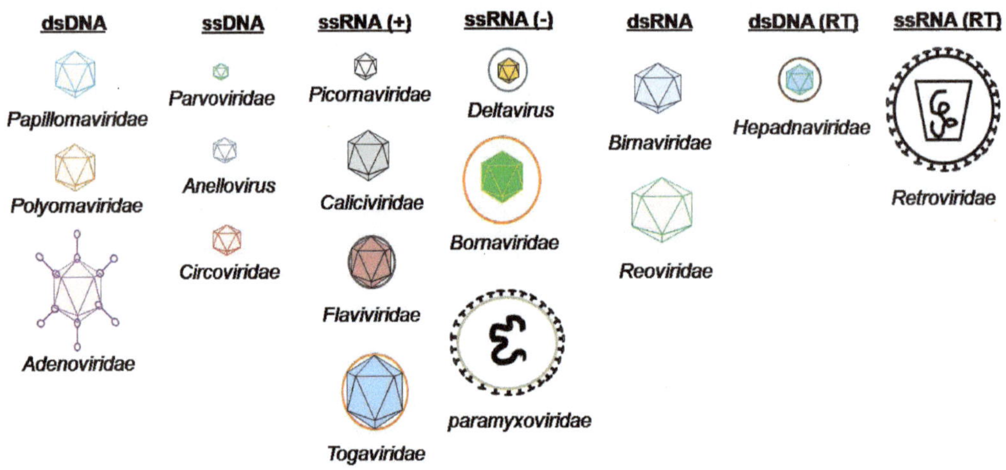

Viral Culture

Viral culture is a laboratory test in which samples are placed with a cell type that the virus being tested for is able to infect. If the cells show changes, known as cytopathic effects, then the culture is positive.

Traditional viral culture has been generally superseded by shell vial culture, in which the sample is centrifuged onto a single layer of cells and viral growth is measured by antigen detection methods. This greatly reduces the time to detection for slow growing viruses such as cytomegalovirus, for which the method was developed. In addition, the centrifugation step in shell viral culture enhances the sensitivity of this method because after centrifugation, the viral particles of the sample are in close proximity to the cells.

Human and monkey cells are used in both traditional viral culture and shell vial culture.

Human virus types that can be identified by viral culture include adenovirus, cytomegalovirus, enteroviruses, herpes simplex virus, influenza virus, parainfluenza virus, rhinovirus, respiratory syncytial virus, varicella zoster virus, measles and mumps. For these, the final identification method is generally by immunofluorescence, with exception of cytomegalovirus and rhinovirus, whose identification in a viral culture are determined by cytopathic effects.

Virus Isolation

Viruses are obligate intracellular parasites that require living cells in order to replicate. Generally cell culture, embryonated eggs and small laboratory animals are used for the isolation of viruses. Embryonated eggs are very useful for the isolation of influenza and paramyxoviruses. Although laboratory animals are useful in isolating different kind of viruses, cell culture is still a preferred way for virus isolation in many of the laboratories.

For primary cell cultures, tissue fragments are first dissociated into small pieces with the help of scissors and addition of trypsin. The cell suspension is then washed couple of times with minimal essential media and seeded into a flat-bottomed glass or plastic container bottle after resuspending it with a suitable liquid medium and fetal calf serum. The cells are kept in incubator at 37°C for 24 to 48hrs depending on the cell type. This allows the cells to attach the surface of the container and its division following the normal cell cycle.

Cell cultures are generally of 3 types:-

1. Primary culture – These are prepared directly from animal or human tissues and can be subcultured only once or twice e.g. chicken embryo fibroblast.

2. Diploid cell culture – They are derived from neonatal tissues and can be subcultured 5-10 times. e.g. human diploid fibroblasts cells.

3. Continuous cells – They are derived from tumor tissues and can be subcultured more than 10 times. e.g. Vero, Hep2, Hela.

Specimens containing virus should be transported to the laboratory as soon as possible upon being taken. Oral or cloacal swabs should be collected in vials containing virus transport medium. Body fluids and tissues should be collected in a sterile container and sealed properly. If possible all the samples should be maintained and transported in a cold condition for higher recovery rates.

Upon receipt, the samples should be inoculated into cell culture depending on the history and symptoms of the disease. The infected cell culture flask should be observed every day for any presence of cytopathic effect (CPE). Certain kind of samples, such as faeces and urine are toxic to the cell cultures and may produce a CPE-like effect. When virus specific CPE is evident, it is advised to passage the infected culture fluid into a fresh cell culture. For cell-associated viruses such as cytomegaloviruses, it is required to trypsinize and passage the intact infected cells. Viruses such as adenovirus can be subcultured after couple of time freezing and thawing of the infected cells.

- Susceptible cell lines:
- Influenza virus- MDCK cells, Vero cells.
- Paramyxoviruses- DF-1 cells, Vero cells.
- Adenoviruses- HEK cells, HuH7 cells.
- Herpesviruses- LMH cells.
- Respiratory syncytia virus- Hep2 cells, Vero cells.
- Virus induced CPE in cell culture

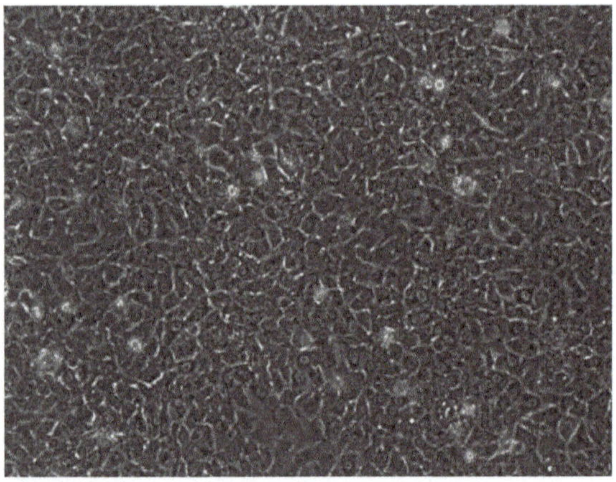

Normal human epithelial cells

Adenovirus infected human epithelial cells

Purification of Virus and Components:

Ultracentrifugation:

The viruses are usually purified with the help of ultracentrifugation. The machine is capable of rotating the samples at 20,000-100,000 rpm under the density gradient of $CsCl_2$ or sucrose. Density at which viruses neither sink nor float when suspended in a density gradient is called as buoyant density . The rate at which viral particles sediment under a defined gravitational force is called as sedimentation coefficient . The basic unit is the Svedberg (S) which is 10 -13 sec. The S value of a virus is used to estimate its molecular weight.

Types of Sedimentation Medium:

A. Sucrose cushions or gradient - A fixed concentration or a linear gradient of sucrose is used. Increasing the density and viscosity of the medium decreases the rate at which virus sediments through them. In g eneral a "cushion" of sucrose is prepared at the bottom of the centrifuge tube and the sample containing virus is overlaid over the cushion. Since most viruses have greater densities than sucrose, separation is based on S values. This method can be used to separate molecules with relatively close S values. Sometime glycerol is also used in place of sucrose.

B. CsCl 2 gradient centrifugation - A linear gradient of CsCl 2 in buffer is prepared in the ultracentrifuge tube. As the concentration of the CsCl 2 is increased the density of the medium increases in the tube so that density is low at the top and high at the bottom. Viral particle centrifuged through this medium will form a band at a position equal to their buoyant density. These are useful to separate viruses of different densities. Limitation of this method is that CsCl 2 can permanently inactivate some viruses.

Other Techniques for Separation:

Viruses can also be separated by electrophoresis and column chromatography but these are not the preferred way to separate virus while sometimes they are used to separate viral nucleic acids or proteins. Both the methods separate the virus on the basis of charge and/or size. Virus contains a variety of charged macromolecule on its surface which contributes to its electrophoretic mobility or ion-exchange characteristics. Viruses are sometimes ligated with the charged group to be separated by ion exchange chromatography. Molecular sieve chromatography can also be used to purify the viruses where large pores are formed with the help of special agarose through which virus particles can enter.

Purity of Viruses:

Many methods are used to assess the purity of virus. The ratio of UV absorption at 260 and 280 nm during a spectrophotometric analysis (260/280) is a characteristic feature to measure the purity of a virus sample and is dependent on the amount of nucleic acid and protein present in the virion. Serological methods such as enzyme-linked immunosorbent assay (ELISA), radioimmuno precipitation assay (RIPA), western blot, virus neutralization test (VNT), and complement fixation are also used to check the puirity of a virus sample. These methods require antibodies specific to viral proteins that may be monoclonal (single type of antibody specific to a single viral protein) or polyclonal (several different antibodies that may recognize several viral proteins or epitopes). Plaque assay is also performed in order to isolate the single colony from a pool of quasispecies viruses.

A general approach for purifying a virus from tissue culture cells

Infect the virus to a confluent cell culture monolayer

↓

Incubate the infected cells at 37^{0}C at 24-48 hrs

↓

Harvest the cells and its supernatant

↓

Freeze thaw couple of time to release the viral particle from the cells

↓

Ultracentrifuge at high speed under sucrose cushion or gradient

↓

Collect the viral pellet

↓

Check for its purity and quantification

References

- PELLICER, JAUME; FAY, MICHAEL F.; LEITCH, ILIA J. (15 September 2010). "The largest eukaryotic genome of them all?". Botanical Journal of the Linnean Society. 164 (1): 10–15. doi:10.1111/j.1095-8339.2010.01072.x
- Ridley, M. (2006), Genome: the autobiography of a species in 23 chapters (PDF), New York: Harper Perennial, ISBN 0-06-019497-9
- Zimmer, Carl (December 18, 2013). "Toe Fossil Provides Complete Neanderthal Genome". New York Times. Retrieved 18 December 2013
- Ellis LL; Huang W; Quinn AM; et al. (2014). "Intrapopulation Genome Size Variation in "Drosophila melanogaster" Reflects Life History Variation and Plasticity". PLoS Genetics. 10 (7): e1004522. PMC 4109859 . PMID 25057905. doi:10.1371/journal.pgen.1004522. Retrieved 17 March 2016
- Griffiths JF; Gelbart WM; Lewontin RC; Wessler SR; Suzuki DT; Miller JH (2005). Introduction to Genetic Analysis. New York: W.H. Freeman and Co. pp. 34–40, 473–476, 626–629. ISBN 0-7167-4939-4
- Wade, Nicholas (2007-05-31). "Genome of DNA Pioneer Is Deciphered". The New York Times. Retrieved 2 April 2010
- Naclerio, G; Cangiano, G; Coulson, A; Levitt, A; Ruvolo, V; La Volpe, A (1992-07-05). "Molecular and genomic organization of clusters of repetitive DNA sequences in Caenorhabditis elegans". Journal of Molecular Biology. 226 (1): 159–68. PMID 1619649. doi:10.1016/0022-2836(92)90131-3
- Lewin, Benjamin (2004). Genes VIII (8th ed.). Upper Saddle River, NJ: Pearson/Prentice Hall. ISBN 0-13-143981-2

- Koonin, Eugene V. (2011-08-31). The Logic of Chance: The Nature and Origin of Biological Evolution. FT Press. ISBN 9780132542494
- Kaplan, Sarah (2016-04-17). "How do your 20,000 genes determine so many wildly different traits? They multitask.". The Washington Post. Retrieved 2016-08-27
- Challacombe, Jean F.; Eichorst, Stephanie A.; Hauser, Loren; Land, Miriam; Xie, Gary; Kuske, Cheryl R.; Steinke, Dirk (15 September 2011). Steinke, Dirk, ed PLoS ONE. 6 (9): e24882. Bibcode:2011PLoSO...624882C. PMC 3174227 . PMID 21949776. doi:10.1371/journal.pone.0024882
- Mankertz P (2008). "Molecular Biology of Porcine Circoviruses". Animal Viruses: Molecular Biology. Caister Academic Press. ISBN 978-1-904455-22-6
- "Virology – Human Immunodeficiency Virus And Aids, Structure: The Genome And Proteins Of HIV". Pathmicro.med.sc.edu. 2010-07-01. Retrieved 27 January 2011
- Stojanovic, edited by Nikola (2007). Computational genomics : current methods. Wymondham: Horizon Bioscience. ISBN 1-904933-30-0
- Court, D. L.; Oppenheim, A. B.; Adhya, S. L. (2007). "A new look at bacteriophage lambda genetic networks". Journal of Bacteriology. 189 (2): 298–304. PMC 1797383 . PMID 17085553. doi:10.1128/JB.01215-06
- NCBI_user_services (29 March 2004). "Mapping Factsheet". Archived from the original on 19 July 2010. Retrieved 27 January 2011
- Storch, Gregory A.; Bernard N. Fields; David Mahan Knipe; Peter M. Howley (2007). "Diagnostic virology". In David Mahan Knipe, Peter M. Howley. Fields' Virology. 1 (5th ed.). Lippincott Williams & Wilkins. p. 3177. ISBN 0-7817-6060-7
- Sanger, F.; Coulson, A.R.; Hong, G.F.; Hill, D.F.; Petersen, G.B. (1982). "Nucleotide sequence of bacteriophage lambda DNA". Journal of Molecular Biology. 162 (4): 729–73. PMID 6221115. doi:10.1016/0022-2836(82)90546-0

3

Virus Classification and RNA Viruses

Viruses were initially classified on the basis of the diseases they caused or on their geographical location. However, this changed after the introduction of the Baltimore classification. It categorised the viruses on the basis of their genome and replication methods. Classification of viruses and positive strand RNA virus is best understood in confluence with the major topics listed in the following chapter.

Virus Classification

Virus classification is the process of naming viruses and placing them into a taxonomic system. Similar to the classification systems used for cellular organisms, virus classification is the subject of ongoing debate and proposals. This is mainly due to the pseudo-living nature of viruses, which is to say they are non-living particles with some chemical characteristics similar to those of life. As such, they do not fit neatly into the established biological classification system in place for cellular organisms.

Viruses are mainly classified by phenotypic characteristics, such as morphology, nucleic acid type, mode of replication, host organisms, and the type of disease they cause. Currently, two main schemes are used for the classification of viruses: the International Committee on Taxonomy of Viruses (ICTV) system and Baltimore classification system, which places viruses into one of seven groups. Accompanying this broad method of classification are specific naming conventions and further classification guidelines set out by the ICTV.

A catalog of all the world's viruses has been proposed; some related preliminary efforts have been accomplished.

Virus Species Definition

Species form the basis for any biological classification system. The ICTV had adopted the principle that a virus species is a polythetic class of viruses that constitutes a replicating lineage and occupies a particular ecological niche. In July 2013, the ICTV definition of species changed to state: "A species is a monophyletic group of viruses whose properties can be distinguished from those of other species by multiple criteria."

ICTV Classification

The International Committee on Taxonomy of Viruses began to devise and implement rules for the naming and classification of viruses early in the 1970s, an effort that continues to the present. The

ICTV is the only body charged by the International Union of Microbiological Societies with the task of developing, refining, and maintaining a universal virus taxonomy.

The system shares many features with the classification system of cellular organisms, such as taxon structure. However, this system of nomenclature differs from other taxonomic codes on several points. A minor point is that names of orders and families are italicized, unlike in the International Code of Nomenclature for algae, fungi, and plants and International Code of Zoological Nomenclature.

Viral classification starts at the level of order and continues as follows, with the taxon suffixes given in italics:

- Order (*-virales*)
- Family (*-viridae*)
- Subfamily (*-virinae*)
- Genus (*-virus*)
- Species

Species names generally take the form of *[Disease] virus*.

The establishment of an order is based on the inference that the virus families it contains have most likely evolved from a common ancestor. The majority of virus families remain unplaced. As of 2012, seven orders, 96 families, 22 subfamilies, 420 genera, and 2,618 species of viruses have been defined by the ICTV. The orders are the *Caudovirales, Herpesvirales, Ligamenvirales, Mononegavirales, Nidovirales, Picornavirales,* and *Tymovirales*. These orders span viruses with varying host ranges. The *Ligamenvirales*, infecting archaea, are the most recent addition to the classification system.

Caudovirales are tailed dsDNA (group I) bacteriophages.

Herpesvirales contain large eukaryotic dsDNA viruses.

Ligamenvirales contains linear, dsDNA (group I) archaean viruses.

Mononegavirales include nonsegmented (-) strand ssRNA (Group V) plant and animal viruses.

Nidovirales are composed of (+) strand ssRNA (Group IV) viruses with vertebrate hosts.

Picornavirales contains small (+) strand ssRNA viruses that infect a variety of plant, insect and animal hosts.

Tymovirales contain monopartite (+) ssRNA viruses that infect plants.

Other variations occur between the orders: *Nidovirales*, for example, are isolated for their differentiation in expressing structural and nonstructural proteins separately.

Structure-based Virus Classification

It has been suggested that similarity in virion assembly and structure observed for certain viral groups infecting hosts from different domains of life (e.g., bacterial tectiviruses and eukaryotic adenoviruses or prokaryotic Caudovirales and eukaryotic herpesviruses) reflects an evolutionary relationship between these viruses. Therefore, structural relationship between viruses has been suggested to be used as a basis for defining higher-level taxa - structure-based viral lineages - that could complement the existing ICTV classification scheme.

Baltimore Classification

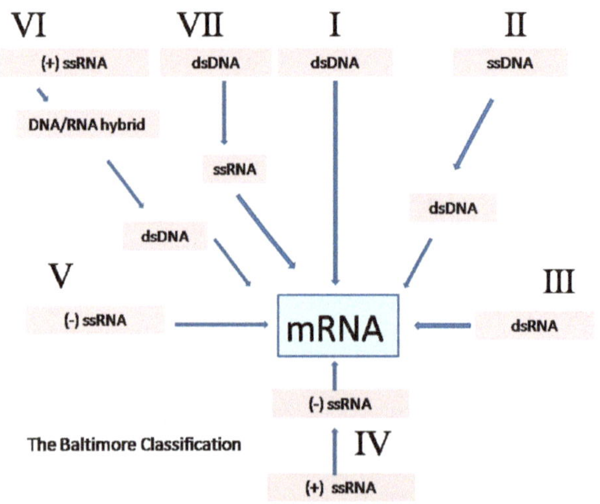

The Baltimore Classification of viruses is based on the method of viral mRNA synthesis

Baltimore classification (first defined in 1971) is a classification system that places viruses into one of seven groups depending on a combination of their nucleic acid (DNA or RNA), strandedness (single-stranded or double-stranded), Sense, and method of replication. Named after David Baltimore, a Nobel Prize-winning biologist, these groups are designated by Roman numerals. Other classifications are determined by the disease caused by the virus or its morphology, neither of which are satisfactory due to different viruses either causing the same disease or looking very similar. In addition, viral structures are often difficult to determine under the microscope. Classifying viruses according to their genome means that those in a given category will all behave in a similar fashion, offering some indication of how to proceed with further research. Viruses can be placed in one of the seven following groups:

- I: dsDNA viruses (e.g. Adenoviruses, Herpesviruses, Poxviruses)
- II: ssDNA viruses (+ strand or "sense") DNA (e.g. Parvoviruses)
- III: dsRNA viruses (e.g. Reoviruses)
- IV: (+)ssRNA viruses (+ strand or sense) RNA (e.g. Picornaviruses, Togaviruses)
- V: (−)ssRNA viruses (− strand or antisense) RNA (e.g. Orthomyxoviruses, Rhabdoviruses)
- VI: ssRNA-RT viruses (+ strand or sense) RNA with DNA intermediate in life-cycle (e.g. Retroviruses)
- VII: dsDNA-RT viruses (e.g. Hepadnaviruses)

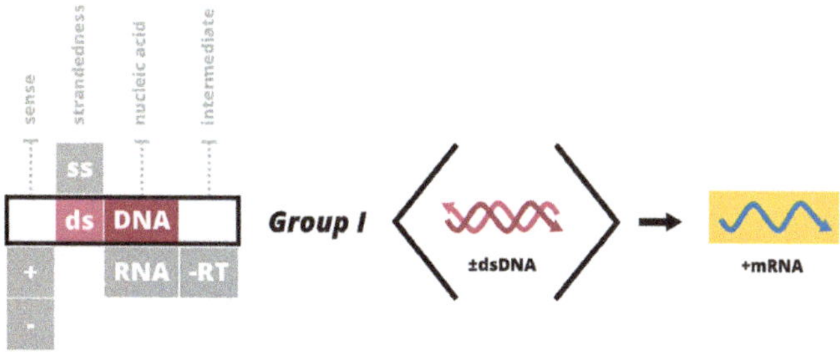

Visualization of the 7 groups of virus according to the Baltimore Classification

DNA Viruses

- Group I: viruses possess double-stranded DNA. Viruses that cause chickenpox and herpes are found here.
- Group II: viruses possess single-stranded DNA.

Virus family	Examples (common names)	Virion naked/enveloped	Capsid symmetry	Nucleic acid type	Group
1. Adenoviridae	Adenovirus, infectious canine hepatitis virus	Naked	Icosahedral	ds	I
2. Papovaviridae	Papillomavirus, polyomaviridae, simian vacuolating virus	Naked	Icosahedral	ds circular	I
3. Parvoviridae	Parvovirus B19, canine parvovirus	Naked	Icosahedral	ss	II
4. Herpesviridae	Herpes simplex virus, varicella-zoster virus, cytomegalovirus, Epstein–Barr virus	Enveloped	Icosahedral	ds	I
5. Poxviridae	Smallpox virus, cow pox virus, sheep pox virus, orf virus, monkey pox virus, vaccinia virus	Complex coats	Complex	ds	I
6. Hepadnaviridae	Hepatitis B virus	Enveloped	Icosahedral	circular, partially ds	VII
7. Anelloviridae	Torque teno virus	Naked	Icosahedral	ss circular	II

RNA Viruses

- Group III: viruses possess double-stranded RNA genomes, e.g. rotavirus.
- Group IV: viruses possess positive-sense single-stranded RNA genomes. Many well known viruses are found in this group, including the picornaviruses (which is a family of viruses that includes well-known viruses like Hepatitis A virus, enteroviruses, rhinoviruses, poliovirus, and foot-and-mouth virus), SARS virus, hepatitis C virus, yellow fever virus, and rubella virus.
- Group V: viruses possess negative-sense single-stranded RNA genomes. The deadly Ebola and Marburg viruses are well known members of this group, along with influenza virus, measles, mumps and rabies.

Virus Family	Examples (common names)	Capsid naked/enveloped	Capsid Symmetry	Nucleic acid type	Group
1. Reoviridae	Reovirus, rotavirus	Naked	Icosahedral	ds	III
2. Picornaviridae	Enterovirus, rhinovirus, hepato-virus, cardiovirus, aphthovirus, poliovirus, parechovirus, erbo-virus, kobuvirus, teschovirus, coxsackie	Naked	Icosahedral	ss	IV
3. Caliciviridae	Norwalk virus	Naked	Icosahedral	ss	IV
4. Togaviridae	Rubella virus, alphavirus	Enveloped	Icosahedral	ss	IV
5. Arenaviridae	Lymphocytic choriomeningitis virus	Enveloped	Complex	ss(-)	V
6. Flaviviridae	Dengue virus, hepatitis C virus, yellow fever virus	Enveloped	Icosahedral	ss	IV
7. Orthomyxoviridae	Influenzavirus A, influenzavirus B, influenzavirus C, isavirus, thogotovirus	Enveloped	Helical	ss(-)	V
8. Paramyxoviridae	Measles virus, mumps virus, re-spiratory syncytial virus, Rinder-pest virus, canine distemper virus	Enveloped	Helical	ss(-)	V
9. Bunyaviridae	California encephalitis virus, hantavirus	Enveloped	Helical	ss(-)	V
10. Rhabdoviridae	Rabies virus	Enveloped	Helical	ss(-)	V
11. Filoviridae	Ebola virus, Marburg virus	Enveloped	Helical	ss(-)	V
12. Coronaviridae	Corona virus	Enveloped	Helical	ss	IV
13. Astroviridae	Astrovirus	Naked	Icosahedral	ss	IV
14. Bornaviridae	Borna disease virus	Enveloped	Helical	ss(-)	V
15. Arteriviridae	Arterivirus, equine arteritis virus	Enveloped	Icosahedral	ss	IV
16. Hepeviridae	Hepatitis E virus	Naked	Icosahedral	ss	IV

Reverse Transcribing Viruses

- Group VI: viruses possess single-stranded RNA viruses that replicate through a DNA intermediate. The retroviruses are included in this group, of which HIV is a member.
- Group VII: viruses possess double-stranded DNA genomes and replicate using reverse transcriptase. The hepatitis B virus can be found in this group.

Holmes Classification

Holmes (1948) used Carl Linnaeus's system of binomial nomenclature to classify viruses into 3 groups under one order, Virales. They are placed as follows:

- Group I: *Phaginae* (attacks bacteria)
- Group II: *Phytophaginae* (attacks plants)
- Group III: *Zoophaginae* (attacks animals)

LHT System of Virus Classification

The LHT System of Virus Classification is based on chemical and physical characters like nucleic acid (DNA or RNA), Symmetry (Helical or Icosahedral or Complex), presence of envelope, diameter of capsid, number of capsomers. This classification was approved by the Provisional Committee on Nomenclature of Virus (PNVC) of the International Association of Microbiological Societies (1962). It is as follows:

- Phylum Vira (divided into 2 subphyla)
  - Subphylum Deoxyvira (DNA viruses)
    - Class Deoxybinala (dual symmetry)
      - Order Urovirales
        - Family Phagoviridae
    - Class Deoxyhelica (Helical symmetry)
      - Order Chitovirales
        - Family Poxviridae
    - Class Deoxycubica (cubical symmetry)
      - Order Peplovirales
        - Family Herpesviridae (162 capsomeres)
      - Order Haplovirales (no envelope)
        - Family Iridoviridae (812 capsomeres)
        - Family Adenoviridae (252 capsomeres)
        - Family Papiloviridae (72 capsomeres)
        - Family Paroviridae (32 capsomeres)
        - Family Microviridae (12 capsomeres)
  - Subphylum Ribovira (RNA viruses)
    - Class Ribocubica
      - Order Togovirales
        - Family Arboviridae
      - Order Lymovirales
        - Family Napoviridae

            - Family Reoviridae
    - Class Ribohelica
        - Order Sagovirales
            - Family Stomataviridae
            - Family Paramyxoviridae
            - Family Myxoviridae
        - Order Rhabdovirales
            - Suborder Flexiviridales
                - Family Mesoviridae
                - Family Peptoviridae
            - Suborder Rigidovirales
                - Family Pachyviridae
                - Family Protoviridae
                - Family Polichoviridae

Subviral Agents

The following agents are smaller than viruses but have only some of their properties.

Viroids

- Family *Avsunviroidae*
    - Genus *Avsunviroid*; type species: *Avocado sunblotch viroid*
    - Genus *Pelamoviroid*; type species: *Peach latent mosaic viroid*
    - Genus *Elaviroid*; type species: *Eggplant latent viroid*
- Family *Pospiviroidae*
    - Genus *Pospiviroid*; type species: *Potato spindle tuber viroid*
    - Genus *Hostuviroid*; type species: *Hop stunt viroid*
    - Genus *Cocadviroid*; type species: *Coconut cadang-cadang viroid*
    - Genus *Apscaviroid*; type species: *Apple scar skin viroid*
    - Genus *Coleviroid*; type species: *Coleus blumei viroid 1*

Satellites

Satellites depend on co-infection of a host cell with a helper virus for productive multiplication. Their nucleic acids have substantially distinct nucleotide sequences from either their helper virus or host. When a satellite subviral agent encodes the coat protein in which it is encapsulated, it is then called a satellite virus.

- Satellite viruses
  - Single-stranded RNA satellite viruses
    - Subgroup 1: *Chronic bee-paralysis satellite virus*
    - Subgroup 2: *Tobacco necrosis satellite virus*
  - Double-stranded DNA satellite viruses (virophages)
- Satellite nucleic acids
  - Single-stranded satellite DNAs
  - Double-stranded satellite RNAs
  - Single-stranded satellite RNAs
    - Subgroup 1: Large satellite RNAs
    - Subgroup 2: Small linear satellite RNAs
    - Subgroup 3: Circular satellite RNAs (virusoids)

Prions

Prions, named for their description as "*pro*teinaceous and *in*fectious particles", lack any detectable (as of 2002) nucleic acids or virus-like particles. They resist inactivation procedures that normally affect nucleic acids.

- Mammalian prions:
  - Agents of spongiform encephalopathies
- Fungal prions:
  - PSI+ prion of *Saccharomyces cerevisiae*
  - URE3 prion of *Saccharomyces cerevisiae*
  - RNQ/PIN+ prion of *Saccharomyces cerevisiae*
  - Het-s prion of *Podospora anserina*

Defective Interfering Particles

- Defective interfering RNA
- Defective interfering DNA

Baltimore Classification

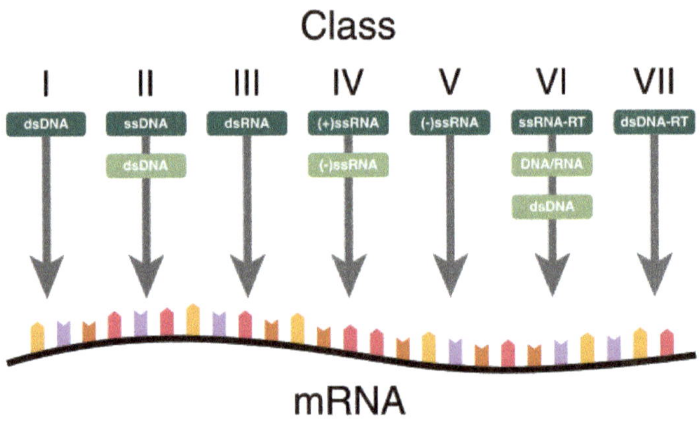

The Baltimore Classification of viruses, created by the American biologist David Baltimore, is based on the method of viral mRNA synthesis

The Baltimore classification, developed by David Baltimore, is a virus classification system that groups viruses into families, depending on their type of genome (DNA, RNA, single-stranded (ss), double-stranded (ds), etc..) and their method of replication.

Classifications

Classifying viruses according to their genome means that those in a given category will all behave in much the same way, which offers some indication of how to proceed with further research. In short:

- I: dsDNA viruses (e.g. Adenoviruses, Herpesviruses, Poxviruses)
- II: ssDNA viruses (+ strand or "sense") DNA (e.g. Parvoviruses)
- III: dsRNA viruses (e.g. Reoviruses)
- IV: (+)ssRNA viruses (+ strand or sense) RNA (e.g. Picornaviruses, Togaviruses)
- V: (−)ssRNA viruses (− strand or antisense) RNA (e.g. Orthomyxoviruses, Rhabdoviruses)
- VI: ssRNA-RT viruses (+ strand or sense) RNA with DNA intermediate in life-cycle (e.g. Retroviruses)
- VII: dsDNA-RT viruses (e.g. Hepadnaviruses)

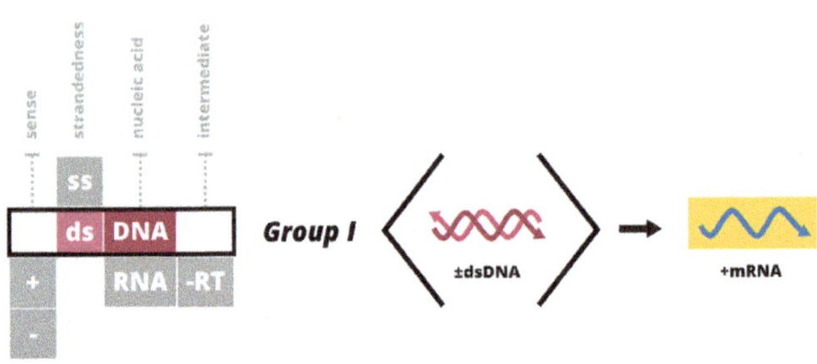

Visualization of the 7 groups of virus according to the Baltimore Classification

Group I: Double-stranded DNA Viruses

These types of viruses must enter the host nucleus before they are able to replicate. Furthermore, these viruses require host cell polymerases to replicate the viral genome and, hence, are highly dependent on the cell cycle. Proper infection and production of progeny requires that the cell be in replication, as it is during replication that the cell's polymerases are active. The virus may induce the cell to forcefully undergo cell division, which may lead to transformation of the cell and, ultimately, cancer. Examples include Herpesviridae, Adenoviridae, and Papovaviridae.

There is only one well-studied example in which a class 1 virus is not replicating within the nucleus: the Poxvirus family, a highly pathogenic virus that infects vertebrates and includes the smallpox virus.

The mRNA is transcribed in the normal way from viral DNA using the host transcriptase enzymes, into two types of mRNA's: 1) early mRNA, transcribed prior to the synthesis of viral DNA, and 2) late mRNA, transcribed from progeny DNA.

Group II: Single-stranded DNA Viruses

Viruses in this category include the Anelloviridae, Circoviridae, and Parvoviridae (which infect vertebrates), the Geminiviridae and Nanoviridae (which infect plants), and the Microviridae (which infect prokaryotes). Most of them have circular genomes (the parvoviruses are the only known exception). Eukaryote-infecting viruses replicate mostly within the nucleus - usually via a rolling circle mechanism, forming double-stranded DNA intermediate in the process. A prevalent but asymptomatic human Anellovirus, called Transfusion Transmitted Virus (TTV), is included within this classification.

Group III: Double-stranded RNA Viruses

As with most RNA viruses, this class replicates in the "Core" capsid that is in cytoplasm, not having to use the host replication polymerases to as much a degree as DNA viruses. This family is also not as well-studied as the rest and includes 2 major families, the Reoviridae and Birnaviridae. Replication is monocistronic and includes individual, segmented genomes, meaning that each of the genes codes for only one protein, unlike other viruses that exhibit more complex translation.

Group IV & V: Single-stranded RNA Viruses

The ssRNA viruses belong to Class IV or V of the Baltimore classification. They could be grouped into positive sense or negative sense according to the sense of polarity of RNA. The single stranded RNA is the common feature of these viruses. The replication of viruses happens in the cytoplasm or nucleus (for segmented class V viruses). Class IV and V ssRNA viruses do not depend as heavily as DNA viruses on the cell cycle.

Group IV: Single-stranded RNA viruses - Positive-sense

The positive-sense RNA viruses and indeed all RNA defined as positive-sense can be directly accessed by host ribosomes to immediately form proteins. These can be divided into two groups, both of which reproduce in the cytoplasm:

- Viruses with polycistronic mRNA where the genome RNA forms the mRNA and is translated into a polyprotein product that is subsequently cleaved to form the mature proteins. This means that the gene can utilize a few methods in which to produce proteins from the same strand of RNA, all in the sake of reducing the size of its gene.
- Viruses with complex transcription, for which subgenomic mRNAs, ribosomal frameshifting, and proteolytic processing of polyproteins may be used. All of which are different mechanisms with which to produce proteins from the same strand of RNA.

Examples of this class include the families Astroviridae, Caliciviridae, Coronaviridae, Flaviviridae, Picornaviridae, Arteriviridae, and Togaviridae.

Group V: Single-stranded RNA Viruses - Negative-sense

The negative-sense RNA viruses and indeed all genes defined as negative-sense cannot be directly accessed by host ribosomes to immediately form proteins. Instead, they must be transcribed by viral polymerases into a "readable" form, which is the positive-sense reciprocal. These can also be divided into two groups:

- Viruses containing nonsegmented genomes for which the first step in replication is transcription from the (-)-stranded genome by the viral RNA-dependent RNA polymerase to yield monocistronic mRNAs that code for the various viral proteins. A positive-sense genome copy is then produced that serves as template for production of the (-)-strand genome. Replication is within the cytoplasm.
- Viruses with segmented genomes for which replication occurs in the nucleus and for which the viral RNA-dependent RNA polymerase produces monocistronic mRNAs from each genome segment. The largest difference between the two is the location of replication.

Examples in this class include the families Arenaviridae, Orthomyxoviridae, Paramyxoviridae, Bunyaviridae, Filoviridae, and Rhabdoviridae (the latter of which includes the rabies virus).

Group VI: Positive-sense Single-stranded RNA Viruses that Replicate through a DNA Intermediate

A well-studied family of this class of viruses include the retroviruses. One defining feature is the use of reverse transcriptase to convert the positive-sense RNA into DNA. Instead of using the RNA for templates of proteins, they use DNA to create the templates, which is spliced into the host genome using integrase. Replication can then commence with the help of the host cell's polymerases.

Group VII: Double-stranded DNA Viruses that Replicate through a Single-stranded RNA Intermediate

This small group of viruses, exemplified by the Hepatitis B virus (which is in the Hepadnaviridae family), have a double-stranded, gapped genome that is subsequently filled in to form a covalently closed circle (cccDNA) that serves as a template for production of viral mRNAs and a subgenomic RNA. The pregenome RNA serves as template for the viral reverse transcriptase for production of the DNA genome.

Non Approved Specifications

Some specification not approved by the ICTV:

Strain - These are different lines of isolates of the same virus. Eg. Influenza viruses those were isolated from different geographical locations.

Type - They show different reactivity towards a positive serum sample, sometime called as sero-types (different antigenic specificity) of the same virus. Eg. Paramyxovirus type 1-9. There may also be subtypes within a particular type.

Group - These are divisions often based on nucleotide sequence similarities or origin. HIV group M (Main), N (Neither M or O), or O (Outlier). There may also be subgroups. (also called clade) within a particular group (M group HIV has A-J subgroups).

Variant - These are viruses whose phenotype differs from original wild type strain.

Origins of Some Viral Names

Picorna: small having size in the scale of 10^{-12} RNA segment

Birna: two RNA segment

Toga: wearing a robe

Rota : Wheel like

Arbo- Arthopod borne

Papilloma: infections result in warts

Adeno: infections of glands

Hepadna: hepatitis + DNA

Herp es: produce scaly lesions

Pox: p roduce pox lesions

Corona: crown like

Satellite Viruses and Defective Interfering particles:

Consider viruses to be a part of ecological habitat where organisms tend to share the relationships with one another: mutualism, commensalism, symbiosis, and parasitism. Viruses also act similarly .

Satellite viruses - Viruses with separate genomes that are encapsidated inside viral particles that are produced by a "helper" virus. They also require helper virus replicative machinery to replicate their genomes.

Defective Interfering particles (DI particles) - Their genomes are derived from a helper virus. They are deletion mutants which have lost their ability to encode proteins, but retain their ability to replicate with the help of a replication machinery of other helper virus. They called defective interfering particles because they are defective in their a bility to produce proteins, and tend to interfere with the replication of helper virus by competing with the resources .

Principles of Virus Nomenclature

Followings are the principles of virus nomenclature set by ICTV

- It aims to provide stability, avoid confusion, and to avoid creation of unnecessary names.
- It is independent of any other biological nomenclature.
- The basic unit of classification is TAXON .
- A taxon should be officially approved by the ICTV committee.

The universal virus nomenclature should follow a hierarchical level of Order, Family, Subfamily, Genus, and Species.

Current Status of Virus Taxonomy:

Order	Numbers of Families
Caudovirales	3
Herpesvirales	3
Mononegavirales	4
Nidovirales	3
Picornavirales	5
Tymovirales	4
Not assigned any order	72

Classification of Viruses based on the Type of Nucleic Acid:

dsDNA viruses	ssDNA viruses	dsRNA viruses	ssRNA (+) viruses	ssRNA (-) viruses	RNA and DNA (RT) viruses
Poxviridae	*Circoviridae*	*Reoviridae*	*Picornaviridae*	*Bornaviridae*	*Retroviridae* (RNA)
Asfaviridae	*Anellovirus*	*Birnaviridae*	*Caliciviridae*	*Rhabdoviridae*	*Hepadnaviridae*
Iridoviridae	*Parvoviridae*		*Hepevirus*	*Filoviridae*	(DNA)
Herpesviridae			*Astroviridae*	*Paramyxoviridae*	
Adenoviridae			*Nodaviridae*	*Orthomyxoviridae*	
Polyomaviridae			*Coronaviridae*	*Bunyaviridae*	
Papillomaviridae			*Arteriviridae*	*Arenaviridae*	
			Flaviviridae	*Deltavirus*	
			Togaviridae		

Classification of Order Mononegavirales :

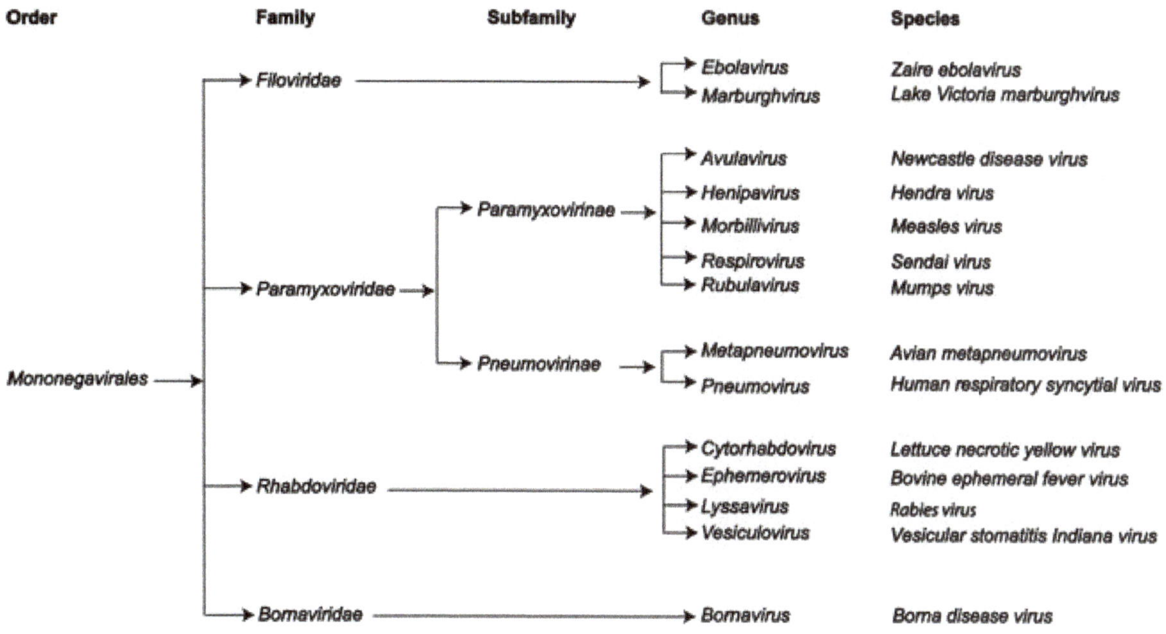

RNA Virus

An RNA virus is a virus that has RNA (ribonucleic acid) as its genetic material. This nucleic acid is usually single-stranded RNA (ssRNA) but may be double-stranded RNA (dsRNA). Notable human diseases caused by RNA viruses include Ebola hemorrhoragic fever, SARS, the common cold, influenza, hepatitis C, West Nile fever, polio and measles.

The ICTV classifies RNA viruses as those that belong to *Group III*, *Group IV* or *Group V* of the Baltimore classification system of classifying viruses and does not consider viruses with DNA intermediates in their life cycle as RNA viruses. Viruses with RNA as their genetic material but that include DNA intermediates in their replication cycle are called retroviruses, and comprise *Group VI* of the Baltimore classification. Notable human retroviruses include HIV-1 and HIV-2, the cause of the disease AIDS.

Another term for RNA viruses that explicitly excludes retroviruses is ribovirus.

Characteristics

Single-stranded RNA Viruses and RNA Sense

RNA viruses can be further classified according to the sense or polarity of their RNA into negative-sense and positive-sense, or ambisense RNA viruses. Positive-sense viral RNA is similar to mRNA and thus can be immediately translated by the host cell. Negative-sense viral RNA is complementary to mRNA and thus must be converted to positive-sense RNA by a RNA polymerase before translation. As such, purified RNA of a positive-sense virus can directly cause infection though it may be less infectious than the whole virus particle. Purified RNA of a negative-sense virus is not infectious by itself as it needs to be transcribed into positive-sense RNA; each virion can be transcribed to several positive-sense RNAs. Ambisense RNA viruses resemble negative-sense RNA viruses, except they also translate genes from the positive strand.

Double-stranded RNA Viruses

The double-stranded (ds)RNA viruses represent a diverse group of viruses that vary widely in host range (humans, animals, plants, fungi, and bacteria), genome segment number (one to twelve), and virion organization (T-number, capsid layers, or turrets). Members of this group include the rotaviruses, renowned globally as the most common cause of gastroenteritis in young children, and picobirnaviruses, renowned worldwide as the most commonly occurring virus in fecal samples of both humans and animals with or without signs of diarrhea. Picobirnaviruses have also been recently reported in respiratory tract samples of pigs and bluetongue virus, an economically important pathogen of cattle and sheep. In recent years, remarkable progress has been made in determining, at atomic and subnanometeric levels, the structures of a number of key viral proteins and of the virion capsids of several dsRNA viruses, highlighting the significant parallels in the structure and replicative processes of many of these viruses.

Mutation Rates

RNA viruses generally have very high mutation rates compared to DNA viruses, because viral RNA polymerases lack the proof-reading ability of DNA polymerases. This is one reason why it is difficult to make effective vaccines to prevent diseases caused by RNA viruses. Retroviruses also have a high mutation rate even though their DNA intermediate integrates into the host genome (and is thus subject to host DNA proofreading once integrated), because errors during reverse transcription are embedded into both strands of DNA before integration. Some genes of RNA virus are important to the viral replication cycles and mutations are not tolerated. For example, the region of the hepatitis C virus genome that encodes the core protein is highly conserved, because it contains an RNA structure involved in an internal ribosome entry site.

Replication

Animal RNA viruses are classified into three distinct groups depending on their genome and mode of replication (and the numerical groups based on the older Baltimore classification):

- Double-stranded RNA viruses (Group III) contain from one to a dozen different RNA molecules, each coding for one or more viral proteins.
- Positive-sense ssRNA viruses (Group IV) have their genome directly utilized as if it were mRNA, with host ribosomes translating it into a single protein that is modified by host and viral proteins to form the various proteins needed for replication. One of these includes RNA-dependent RNA polymerase (RNA replicase), which copies the viral RNA to form a double-stranded replicative form. In turn this directs the formation of new virions.
- Negative-sense ssRNA viruses (Group V) must have their genome copied by an RNA replicase to form positive-sense RNA. This means that the virus must bring along with it the

RNA replicase enzyme. The positive-sense RNA molecule then acts as viral mRNA, which is translated into proteins by the host ribosomes. The resultant protein goes on to direct the synthesis of new virions, such as capsid proteins and RNA replicase, which is used to produce new negative-sense RNA molecules.

Retroviruses (Group VI) have a single-stranded RNA genome but, in general, are not considered RNA viruses because they use DNA intermediates to replicate. Reverse transcriptase, a viral enzyme that comes from the virus itself after it is uncoated, converts the viral RNA into a complementary strand of DNA, which is copied to produce a double-stranded molecule of viral DNA. After this DNA is integrated into the host genome using the viral enzyme integrase, expression of the encoded genes may lead to the formation of new virions.

Classification

Classification of the RNA viruses has proven to be a difficult problem. This is in part due to the high mutation rates these genomes undergo. Classification is based principally on the type of genome (double-stranded, negative- or positive-single-strand) and gene number and organisation. Currently there are 5 orders and 47 families of RNA viruses recognised. There are also many unassigned species and genera.

Related to but distinct from the RNA viruses are the viroids and the RNA satellite viruses. These are not currently classified as RNA viruses and are described on their own pages.

Positive Strand RNA Viruses

This is the single largest group of RNA viruses with 30 families. Attempts have been made to group these families in higher orders. These proposals were based on an analysis of the RNA polymerases and are still under consideration. To date, the suggestions proposed have not been broadly accepted because of doubts over the suitability of a single gene to determine the taxonomy of the clade.

The proposed classification of positive-strand RNA viruses is based on the RNA-dependent RNA polymerase. Three groups have been recognised:

I. Bymoviruses, comoviruses, nepoviruses, nodaviruses, picornaviruses, potyviruses, sobemoviruses and a subset of luteoviruses (beet western yellows virus and potato leafroll virus)—the picorna like group (Picornavirata).

II. Carmoviruses, dianthoviruses, flaviviruses, pestiviruses, statoviruses, tombusviruses, single-stranded RNA bacteriophages, hepatitis C virus and a subset of luteoviruses (barley yellow dwarf virus)—the flavi like group (Flavivirata).

III. Alphaviruses, carlaviruses, furoviruses, hordeiviruses, potexviruses, rubiviruses, tobraviruses, tricornaviruses, tymoviruses, apple chlorotic leaf spot virus, beet yellows virus and hepatitis E virus—the alpha like group (Rubivirata).

A division of the alpha-like (Sindbis-like) supergroup on the basis of a novel domain located near the N termini of the proteins involved in viral replication has been proposed. The two groups proposed are: the 'altovirus' group (alphaviruses, furoviruses, hepatitis E virus, hordeiviruses, tobamoviruses, tobraviruses, tricornaviruses and probably rubiviruses); and the 'typovirus' group (apple chlorotic leaf spot virus, carlaviruses, potexviruses and tymoviruses).

The alpha like supergroup can be further divided into three clades: the rubi-like, tobamo-like, and tymo-like viruses.

Additional work has identified five groups of positive-stranded RNA viruses containing four, three, three, three, and one order(s), respectively. These fourteen orders contain 31 virus families (including 17 families of plant viruses) and 48 genera (including 30 genera of plant viruses). This analysis suggests that alphaviruses and flaviviruses can be separated into two families—the Togaviridae and Flaviridae, respectively—but suggests that other taxonomic assignments, such as the pestiviruses, hepatitis C virus, rubiviruses, hepatitis E virus, and arteriviruses, may be incorrect. The coronaviruses and toroviruses appear to be distinct families in distinct orders and not distinct genera of the same family as currently classified. The luteoviruses appear to be two families rather than one, and apple chlorotic leaf spot virus appears not to be a closterovirus but a new genus of the Potexviridae.

Evolution

The evolution of the picornaviruses based on an analysis of their RNA polymerases and helicases appears to date to the divergence of the eukaryotes. Their putative ancestors include the bacterial group II retroelements, the family of HtrA proteases and DNA bacteriophages.

Double-stranded RNA Viruses

This analysis also suggests that the dsRNA viruses are not closely related to each other but instead belong to four additional classes—Birnaviridae, Cystoviridae, Partitiviridae, and Reoviridae — and one additional order (Totiviridae) of one of the classes of positive ssRNA viruses in the same subphylum as the positive-strand RNA viruses.

One study has suggested a that there are two large clades: One includes the Caliciviridae, Flaviviridae, and Picornaviridae families and a second that includes the Alphatetraviridae, Birnaviridae and Cystoviridae, Nodaviridae, and Permutotretraviridae families.

Group III—dsRNA Viruses

There are eleven families and a number of unassigned genera and species recognised in this group.

- Family Amalgaviridae
- Family Birnaviridae
- Family Chrysoviridae
- Family Cystoviridae
- Family Endornaviridae
- Family Hypoviridae

- Family Megabirnaviridae
- Family Partitiviridae
- Family Picobirnaviridae
- Family Reoviridae—includes Rotavirus
- Family Totiviridae
- Family Quadriviridae
- Genus Botybirnavirus
- Unassigned species
    - *Botrytis porri RNA virus 1*
    - *Circulifer tenellus virus 1*
    - *Cucurbit yellows associated virus*
    - *Sclerotinia sclerotiorum debilitation-associated virus*
    - *Spissistilus festinus virus 1*

Group IV—Positive-sense ssRNA Viruses

There are three orders and 34 families recognised in this group. In addition, there are a number of unclassified species and genera.

- Order Nidovirales
    - Family Arteriviridae
    - Family Coronaviridae — includes Coronavirus, SARS
    - Family Mesoniviridae
    - Family Roniviridae
- Order Picornavirales
    - Family Dicistroviridae
    - Family Iflaviridae
    - Family Marnaviridae
    - Family Picornaviridae—includes Poliovirus, Rhinovirus (a common cold virus), Hepatitis A virus
    - Family Secoviridae includes subfamily Comovirinae
    - Genus Bacillariornavirus
    - Genus Dicipivirus
    - Genus Labyrnavirus

  - Genus Sequiviridae
  - Species Kelp fly virus
- Order Tymovirales
  - Family Alphaflexiviridae
  - Family Betaflexiviridae
  - Family Gammaflexiviridae
  - Family Tymoviridae
- Unassigned
  - Family Alphatetraviridae
  - Family Alvernaviridae
  - Family Astroviridae
  - Family Barnaviridae
  - Family Benyviridae
  - Family Bromoviridae
  - Family Caliciviridae — includes Norwalk virus
  - Family Carmotetraviridae
  - Family Closteroviridae
  - Family Flaviviridae — includes Yellow fever virus, West Nile virus, Hepatitis C virus, Dengue fever virus, Zika virus
  - Family Fusariviridae
  - Family Hepeviridae
  - Family Leviviridae
  - Family Luteoviridae — includes Barley yellow dwarf virus
  - Family Narnaviridae
  - Family Nodaviridae
  - Family Permutotetraviridae
  - Family Potyviridae
  - Family Statovirus
  - Family Togaviridae — includes Rubella virus, Ross River virus, Sindbis virus, Chikungunya virus
  - Family Tombusviridae

- Family Virgaviridae
- Unassigned genera
  - Genus *Blunervirus*
  - Genus *Cilevirus*
  - Genus *Higrevirus*
  - Genus *Idaeovirus*
  - Genus *Negevirus*
  - Genus *Ourmiavirus*
  - Genus *Polemovirus*
  - Genus *Sinaivirus*
  - Genus *Sobemovirus*
- Unassigned species
  - Acyrthosiphon pisum virus
  - Bastrovirus
  - Blackford virus
  - Blueberry necrotic ring blotch virus
  - Chara australis virus
  - Extra small virus
  - Jingmen tick virus
  - Le Blanc virus
  - Nesidiocoris tenuis virus 1
  - Nylanderia fulva virus 1
  - Orsay virus
  - Plasmopara halstedii virus
  - Rosellinia necatrix fusarivirus 1
  - Santeuil virus
  - Solenopsis invicta virus 2
  - Solenopsis invicta virus 3

- Satellite viruses
  - Family Sarthroviridae

- Genus Albetovirus
- Genus Aumaivirus
- Genus Virtovirus
- Chronic bee paralysis virus
- Papanivirus

Group V—Negative-sense ssRNA Viruses

One order and twelve families are currently recognised in this group. There are also a large number of unassigned species.

- Order *Mononegavirales*
    - Family *Bornaviridae*—includes Borna disease virus
    - Family *Filoviridae*—includes Ebola virus, Marburg virus
    - Family *Mymonaviridae*
    - Family *Nyamiviridae*
    - Family *Paramyxoviridae*—includes Measles virus, Mumps virus, Nipah virus, Hendra virus, and Newcastle disease virus
    - Family *Pneumoviridae*-includes Human respiratory syncytial virus
    - Family *Rhabdoviridae*—includes Rabies virus
    - Family *Sunviridae*
    - Genus *Anphevirus*
    - Genus *Arlivirus*
    - Genus *Chengtivirus*
    - Genus *Crustavirus*
    - Genus *Wastrivirus*
- Unassigned families:
    - Family *Arenaviridae*—includes Lassa virus
    - Family *Bunyaviridae*—includes Hantavirus, Crimean-Congo hemorrhagic fever
    - Family *Ophioviridae*
    - Family *Orthomyxoviridae*—includes Influenza viruses
- Unassigned genera:
    - Genus *Deltavirus*—includes Hepatitis D virus

- Genus *Tenuivirus*
- Genus *Emaravirus*

Notes

The majority of fungal viruses are double-stranded RNA viruses. A small number of positive-strand RNA viruses have been described. One report has suggested the possibility of a negative stranded virus.

Positive Strand RNA Viruses

The viruses which contain positive strand RNA genome act directly as mRNA upon infection to a host cell. Most viruses in this category have icosahedral symmetry and vary approximately 50-150 nm in diameter. The members contain positive sense RNA genome containing either lipid envelope (*Togaviridae* , *Flaviviridae* , and *Coronaviridae*) or devoid of envelope (*Picornaviridae*). They are important because they can cause serious life-threatening diseases including hemorrhagic fever and encephalitis (dengue, yellow fever). Members of genus Alphavirus (Sindbis virus, Semiliki Forest virus, and Equine encephalitis) are transmitted to animal by mosquito bite. Among all Alphaviruses sindbis virus is well studied and understood.

Examples of Positive Strand RNA Viruses:

	Family	Virus example(s)
Animal viruses	*Flaviviridae*	West Nile virus Yellow fever virus Dengue virus bovine viral diarrhea virus classical swine fever Hepatitis C virus
	Coronaviridae	Severe acute respiratory syndrome (SARS) virus Avian infectious bronchitis virus (IBV)
	Togaviridae	Rubella virus Chikungunya virus Semliki Forest virus Sindbis virus Eastern equine encephalitis virus Western equine encephalitis virus Venezuelan equine encephalitis virus Ross River virus
	Picornaviridae	Foot and mouth disease virus Human rhinovirus Encephalomyocarditis virus Polio virus Hepatitis A virus
Plant viruses	*Potyviridae*	Potato virus Y
	Flexiviridae	Potato virus X
	Comoviridae	Cowpea mosaic virus

Virion Properties

Virions are spherical in shape and are having uniform appearance. The diameter of virus varies between 70-100 nm. The enveloped virion particle encircles the icosahedral capsid. Envelope contains spikes of viral glycoprotein which are major antigenic determinants of the virus. The spike glycoproteins are highly variable among strains and also between different serotypes.

Structure of Positive Strand RNA Genome

They are positive sense, single-stranded RNA, vary between 9-12 kb in size, with the exception of coronaviruses. Terminal 5' end of the genome is capped while 3' end is polyadenylated. Viruses with single stranded RNA genome do not require secondary or tertiary fold in their capsid to accommodate its genome. That means they are highly organized and tightly packed. Generally a dimer of coat protein interacts with the 3' end of the RNA, which is essential for the virus replication. This interaction is also needed for the packaging of the genomic RNA inside the virion. The viral genomic RNA is arranged in the icosahedral capsid in various ways in order to neutralize the negative charge of the nucleic acids.

Replication of Positive Strand RNA

The positive strand RNA virus transfers its genome directly to the ribosome and starts translation for the synthesis of viral proteins. Infectious cycle begins with the entry of virus into the cell through endocytosis. The genomic RNA uncoats after getting into the cytoplasm of the infected cells. The RNA is then translated into the viral polyprotein precursors which are later cleaved by proteolysis to form the structural and non structural viral proteins. The structural proteins are involved in the maturation and assembly of the virion while nonstructural proteins act as RNA replicating enzyme for genomic RNA synthesis. Some of the viruses in this class form the subgenomic RNA during replication process (Coronavirus, Caliciviruses, and Togaviruses).

Schematic Representations of togavirus and rubella virus genome:

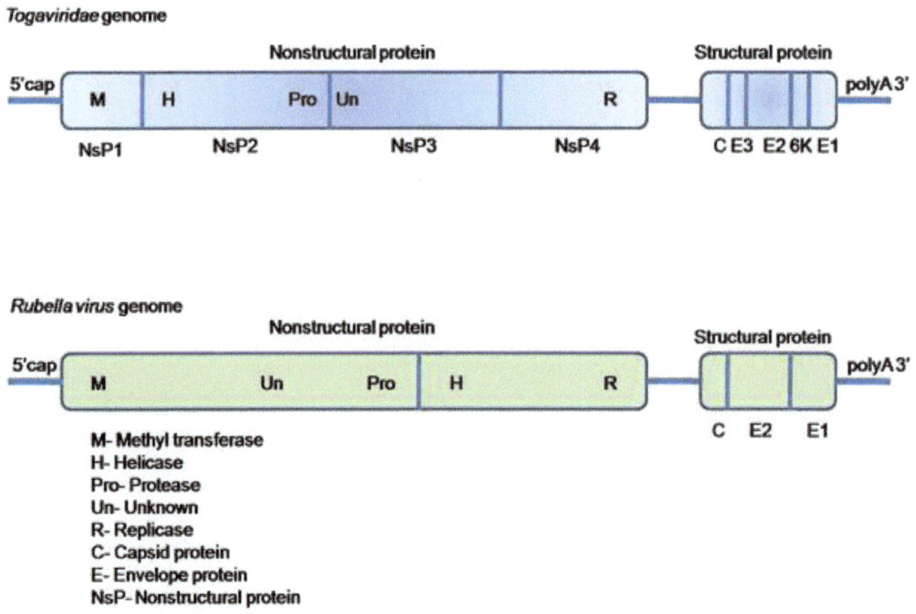

Flavivirus

Flavivirus is a genus of viruses in the family *Flaviviridae*. This genus includes the West Nile virus, dengue virus, tick-borne encephalitis virus, yellow fever virus, Zika virus and several other viruses which may cause encephalitis, as well as insect-specific flaviviruses (ISFs) such as cell fusing agent virus (CFAV), Palm Creek virus (PCV), and Parramatta River virus (PaRV).

Flaviviruses are named from the yellow fever virus, the type virus for the family; the word *flavus* means "yellow" in Latin. The name yellow fever in turn originated from its propensity to cause yellow jaundice in victims.

Flaviviruses share several common aspects: common size (40–65 nm), symmetry (enveloped, icosahedral nucleocapsid), nucleic acid (positive-sense, single-stranded RNA around 10,000–11,000 bases), and appearance in the electron microscope.

Most of these viruses are transmitted by the bite from an infected arthropod (mosquito or tick) and hence, classified as arboviruses. Human infections with these viruses are typically incidental, as humans are usually unable to replicate the virus to high enough titers to reinfect the arthropods needed to continue the virus lifecycle – man is then a dead end host. The exceptions to this are the yellow fever, dengue, and zika viruses, which still require mosquito vectors, but are well-enough adapted to humans as to not necessarily depend upon animal hosts (although they continue to have important animal transmission routes, as well).

Other virus transmission routes for arboviruses include handling infected animal carcasses, blood transfusion, child birth and through consumption of unpasteurised milk products. Transmission from nonhuman vertebrates to humans without an intermediate vector arthropod is thought to be unlikely. For example, early tests with yellow fever showed that the disease is not contagious.

The known non-arboviruses of the flavivirus family reproduce in either arthropods or vertebrates, but not both, with one odd member of the genus affecting a nematode.

Structure

Viruses in Flavivirus are enveloped, with icosahedral and spherical geometries. The diameter is around 50 nm. Genomes are linear positive-sense RNA and non-segmented, around 10-11kb in length.

Genus	Structure	Symmetry	Capsid	Genomic Arrangement	Genomic Segmentation
Flavivirus	Icosahedral-like	Pseudo T=3	Enveloped	Linear	Monopartite

Life Cycle

Entry into the host cell is achieved by attachment of the viral envelope protein E to host receptors, which mediates clathrin-mediated endocytosis. Replication follows the positive stranded RNA virus replication model. Positive stranded RNA virus transcription is the method of transcription.

Humans, mammals, mosquitoes, and ticks serve as the natural host. Transmission routes are zoonosis and bite.

Genus	Host Details	Tissue Tropism	Entry Details	Release Details	Replication Site	Assembly Site	Transmission
Flavivirus	Humans; mammals; mosquitoes; ticks	Epithelium: skin; epithelium: kidney; epithelium: intestine; epithelium: testes	Clathrin-mediated endocytosis	Secretion	Cytoplasm	Cytoplasm	Zoonosis; arthropod bite

Replication

Flaviviruses have a (+) sense RNA genome and replicate in the cytoplasm of the host cells. The genome mimics the cellular mRNA molecule in all aspects except for the absence of the poly-adenylated (poly-A) tail. This feature allows the virus to exploit cellular apparatus to synthesise both structural and non-structural proteins, during replication. The cellular ribosome is crucial to the replication of the flavivirus, as it translates the RNA, in a similar fashion to cellular mRNA, resulting in the synthesis of a single polyprotein. In general, the genome encodes 3 structural proteins (Capsid, prM, and Envelope) and 8 non-structural proteins (NS1, NS2A, NS2B, NS3, NS4A, NS4B, NS5 and NS5B). The genomic RNA is modified at the 5′ end of positive-strand genomic RNA with a cap-1 structure (me^7-GpppA-me^2).

Cellular RNA cap structures are formed via the action of an RNA triphosphatase, with guanylyltransferase, N7-methyltransferase and 2′-O methyltransferase. The virus encodes these activities in its non-structural proteins. The NS3 protein encodes a RNA triphosphatase within its helicase domain. It uses the helicase ATP hydrolysis site to remove the γ-phosphate from the 5′ end of the RNA. The N-terminal domain of the non-structural protein 5 (NS5) has both the N7-methyltransferase and guanylyltransferase activities necessary for forming mature RNA cap structures. RNA binding affinity is reduced by the presence of ATP or GTP and enhanced by S-adenosyl methionine. This protein also encodes a 2′-O methyltransferase.

Once translated, the polyprotein is cleaved by a combination of viral and host proteases to release mature polypeptide products. Nevertheless, cellular post-translational modification is dependent on the presence of a poly-A tail; therefore this process is not host-dependent. Instead, the polyprotein contains an autocatalytic feature which automatically releases the first peptide, a virus specific enzyme. This enzyme is then able to cleave the remaining polyprotein into the individual products. One of the products cleaved is a polymerase, responsible for the synthesis of a (-) sense RNA molecule. Consequently, this molecule acts as the template for the synthesis of the genomic progeny RNA.

Flavivirus genomic RNA replication occurs on rough endoplasmic reticulum membranes in membranous compartments.

New viral particles are subsequently assembled. This occurs during the budding process which is also responsible for the accumulation of the envelope and cell lysis.

A G protein-coupled receptor kinase 2 (also known as ADRBK1) appears to be important in entry and replication for several Flaviviridae.

RNA Secondary Structure Elements

The (+) sense RNA genome of *Flavivirus* contains 5' and 3' untranslated regions (UTRs).

5'UTR

The 5'UTRs are 95–101 nucleotides long in Dengue virus. There are two conserved structural elements in the Flavivirus 5'UTR, a large stem loop (SLA) and a short stem loop (SLB). SLA folds into a Y-shaped structure with a side stem loop and a small top loop. SLA is likely to act as a promoter, and is essential for viral RNA synthesis. SLB is involved in interactions between the 5'UTR and 3'UTR which result in the cyclisation of the viral RNA, which is essential for viral replication.

3'UTR

The 3'UTRs are typically 0.3–0.5 kb in length and contain a number of highly conserved secondary structures which are conserved and restricted to the *flavivirus* family. The majority of analysis has been carried out using West Nile virus (WNV) to study the function the 3'UTR.

Currently 8 secondary structures have been identified within the 3'UTR of WNV and are (in the order in which they are found with the 3'UTR) SL-I, SL-II, SL-III, SL-IV, DB1, DB2 and CRE. Some of these secondary structures have been characterised and are important in facilitating viral replication and protecting the 3'UTR from 5' endonuclease digestion. Nuclease resistance protects the downstream 3' UTR RNA fragment from degradation and is essential for virus-induced cytopathicity and pathogenicity.

- SL-II

SL-II has been suggested to contribute to nuclease resistance. It may be related to another hairpin loop identified in the 5'UTR of the Japanese encephalitis virus (JEV) genome. The JEV hairpin is significantly over-represented upon host cell infection and it has been suggested that the hairpin structure may play a role in regulating RNA synthesis.

- SL-IV

This secondary structure is located within the 3'UTR of the genome of *Flavivirus* upstream of the DB elements. The function of this conserved structure is unknown but is thought to contribute to ribonuclease resistance.

- DB1/DB2

These two conserved secondary structures are also known as pseudo-repeat elements. They were originally identified within the genome of Dengue virus and are found adjacent to each other within the 3'UTR. They appear to be widely conserved across the Flaviviradae. These DB elements have a secondary structure consisting of three helices and they play a role in ensuring efficient translation. Deletion of DB1 has a small but significant reduction in translation but deletion of

DB2 has little effect. Deleting both DB1 and DB2 reduced translation efficiency of the viral genome to 25%.

- CRE

CRE is the Cis-acting replication element, also known as the 3'SL RNA elements, and is thought to be essential in viral replication by facilitating the formation of a "replication complex". Although evidence has been presented for an existence of a pseudoknot structure in this RNA, it does not appear to be well conserved across flaviviruses. Deletions of the 3' UTR of flaviviruses have been shown to be lethal for infectious clones.

Conserved Hairpin cHP

A conserved hairpin (cHP) structure was later found in several *Flavivirus* genomes and is thought to direct translation of capsid proteins. It is located just downstream of the AUG start codon.

The Role of RNA Secondary Structures in sfRNA Production

Subgenomic flavivirus RNA (sfRNA) is an extension of the 3' UTR and has been demonstrated to play a role in flavivirus replication and pathogenesis. sfRNA is produced by incomplete degradation of genomic viral RNA by the host cells 5'-3' exoribonuclease 1 (XRN1). As the XRN1 degrades viral RNA, it stalls at stemloops formed by the secondary structure of the 5' and 3' UTR. This pause results in an undigested fragment of genome RNA known as sfRNA. sfRNA influences the life cycle of the flavivirus in a concentration dependent manner. Accumulation of sfRNA causes (1) antagonization of the cell's innate immune response, thus decreasing host defense against the virus (2) inhibition of XRN1 and Dicer activity to modify RNAi pathways that destroy viral RNA (3) modification of the viral replication complex to increase viral reproduction. Overall, sfRNA is implied in multiple pathways that compromise host defenses and promote infection by flaviviruses.

Evolution

The flaviviruses can be divided into 2 clades: one with the vector borne viruses and the other with no known vector. The vector clade in turn can be subdivided into a mosquito borne clade and a tick borne clade. These groups can be divided again.

The mosquito group can be divided into two branches: one branch contains the neurotropic viruses, often associated with encephalitic disease in humans or livestock. This branch tends to be spread by *Culex* species and to have bird reservoirs. The second branch is the non-neurotropic viruses which are associated with haemorrhagic disease in humans. These tend to have *Aedes* species as vectors and primate hosts.

The tick-borne viruses also form two distinct groups: one is associated with seabirds and the other - the tick-borne encephalitis complex viruses - is associated primarily with rodents.

The viruses that lack a known vector can be divided into three groups: one closely related to the mosquito-borne viruses which is associated with bats; a second, genetically more distant, is also associated with bats; and a third group is associated with rodents.

It seems likely that tick transmission may have been derived from a mosquito borne group.

A partial genome of a flavivirus has been found in the sea spider *Endeis spinosa*. The sequences are related to those in the insect specific flaviviruses. It is not presently clear how this sequence fits into the evolution of this group of viruses.

Properties of Flaviviruses

Virions are spherical and 40-60nm in diameter. They contain a lipid derived envelope with spikes of glycoprotein embedded on it. The genome consists of a positive sense single-stranded RNA of approximately 9.6 to 12.3 kbp. 5' cap is present only in the members of genus *Flavivirus* . The viral genome codes for both structural (3-5 in numbers) as well as non-structural (7-8 in numbers) proteins. The viruses are easily inactivated by common disinfectants and heat. Flaviviruses infect a variety of cells including Vero (African green monkey), BHK-21 (baby hamster kidney) and chicken embryo fibroblasts. Infection of flavivirus is lethal to new born mice. Viruses enter the cells through receptor mediated endocytosis and replication of the viral genome takes place in the cytoplasm. Replication involves synthesis of negative sense RNA from the positive sense genomic RNA, which then serves as a template for mRNA synthesis. Translation of viral mRNA leads to formation of a single polyprotein which is then cleaved into structural and non structural proteins. The maturation of the virion takes place in endoplasmic reticulum and are released following lysis of the infected cells.

West Nile Virus

West Nile virus was first identified in 1937 in Africa as a cause of mild febrile infection. Later on it was identified as a causative agent of fatal encephalitis in humans and horses. The virus is transmitted to the human being by the bite of Culex mosquitoes, while birds serve as a reservoir host. The animal and humans are considered as an incidental host when virus gets transmitted following insect bite harboring the virus in large quantity.

West Nile Virus Transmission Cycle

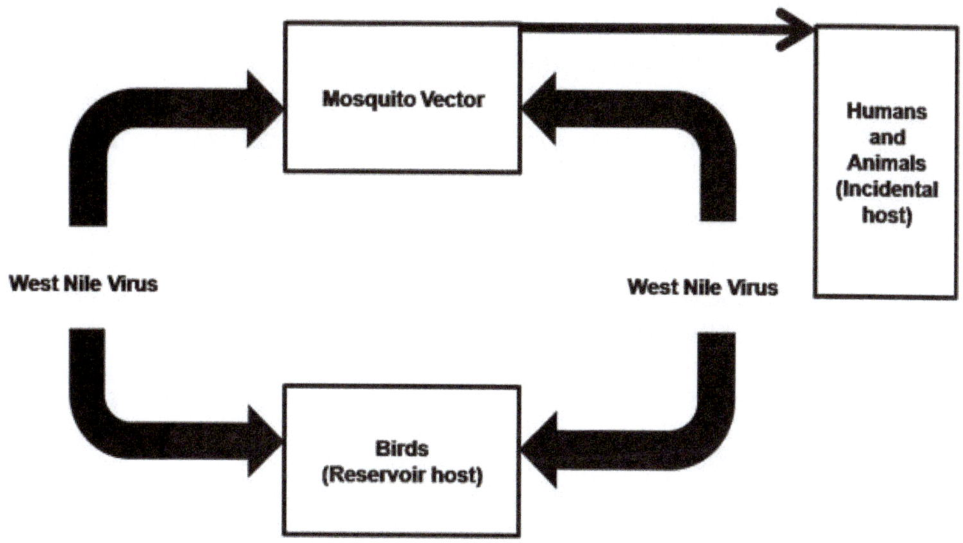

Ecology of West Nile Virus Transmission:

Weather plays an important role in the transmission of West Nile virus to humans and animals. Change in temperature and dryness is an ideal condition for the breeding of mosquitoes. Also during the rainy season the behavioral patterns of birds, mosquitoes, and humans changes. These lead to a hypothesis of having high incidence rate of West Nile virus during initial precipitation and dryness in the air. The role of wind velocity and pressure difference has also been suggested for the outcomes of West Nile virus incidences, but their actual impact is an open area of discussion.

Ecology of West Nile Virus Transmission:

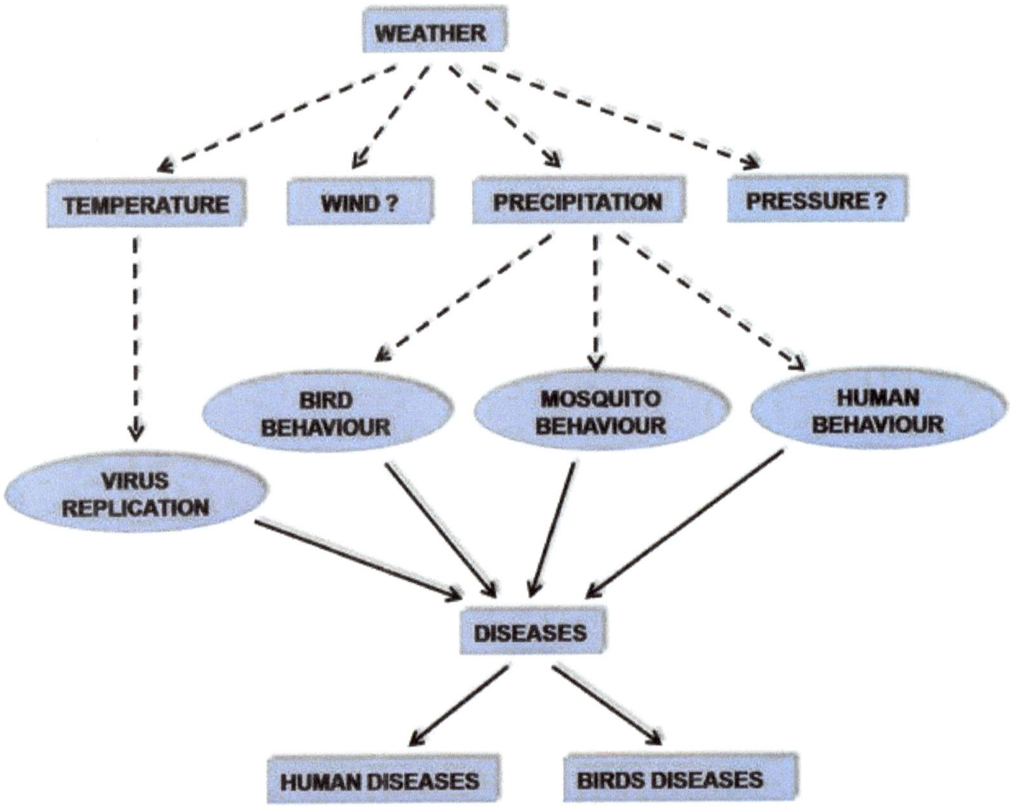

Clinical Features:

Birds often contains virus and do not show any disease symptoms while humans and horses are dead-end hosts. Clinical manifestation of the disease is only evident in case of horses and humans. Infected animals show neurological signs, depression, muscle weakness, and fever. Death occurs in 40% of infected cases based on the immune status of the animal and the strain of virus. Milder form of disease in humans shows fever, abdominal pain, diarrhea and restlessness. These symptoms last for a week. More severe form that can be life threatening is called as West Nile encephalitis.

Pathogenesis:

Infection of West Nile virus causes high-titer viremia, necrosis, hemorrhage and inflammation in many vital organs. These include heart, brain, liver, kidney, intestine, and nervous system. Lesions are evident on brain and spinal cord following West Nile virus infection.

Diagnosis:

Diagnosis of this disease is done mainly on the basis of serology for virus specific immunoglobulin in the serum of infected patient. Traditional techniques such as virus neutralization assay, *in–situ* hybridization, and immuno-histochemistry are regularly used to diagnose the disease. Modern molecular biological techniques such as reverse transcriptase polymerase chain reaction (RT-PCR) are also used to detect the virus infection from tissue samples.

Immunity, Prevention and Control:

Animal previously infected with West Nile virus are resistant to reinfection and many vaccines are available in the market against West Nile disease. The possible way to control the disease is by controlling mosquito population.

Dengue Virus

Dengue virus (DENV) is the cause of dengue fever. It is a mosquito-borne single positive-stranded RNA virus of the family *Flaviviridae*; genus *Flavivirus*. Five serotypes of the virus have been found, all of which can cause the full spectrum of disease. Nevertheless, scientists are finding their understanding of dengue virus may be simplistic, as rather than distinct antigenic groups there appears to be a *continuum*. This same study identified 47 strains of dengue virus. Additionally, coinfection with and lack of rapid tests for zika virus and chikungunya complicate matters in real world infections.

Its genome is about 11000 bases of positive-sense single stranded RNA (ssRNA) that codes for three structural proteins (capsid protein C, membrane protein M, envelope protein E) and seven nonstructural proteins (NS1, NS2a, NS2b, NS3, NS4a, NS4b, NS5). It also includes short non-coding regions on both the 5' and 3' ends.

Evolution

The dengue type 1 virus appears to have evolved in the early 19th century. Based on the analysis of the envelope protein there are at least four genotypes (1 to 4). In 2013 a fifth serotype was reported. The rate of nucleotide substitution for this virus has been estimated to be 6.5×10^{-4} per nucleotide per year, a rate similar to other RNA viruses. The American African genotype has been estimated to have evolved between 1907 and 1949. This period includes World Wars I and II, which were associated with considerable movement of populations and environmental disturbance, factors known to promote the evolution of new vector-borne viral species.

Life Cycle

Until a few hundred years ago dengue virus was transmitted in sylvatic cycles in Africa and Asia between mosquitoes of the genus *Aedes* and non-human primates with rare emergences into human populations. The global spread of dengue virus, however, has followed its emergence from sylvatic cycles and the primary life cycle now exclusively involves transmission between humans and *Aedes* mosquitoes. Vertical transmission from mosquito to mosquito has also been observed in some vector species.

As the virus infects human cells, recent findings suggest that host homeostatic processes like autophagy and ER stress response not to mention apoptosis are triggered depending on the infected cell type. The activation of autophagy and ER stress during infection enhances virus reproduction.

Structures

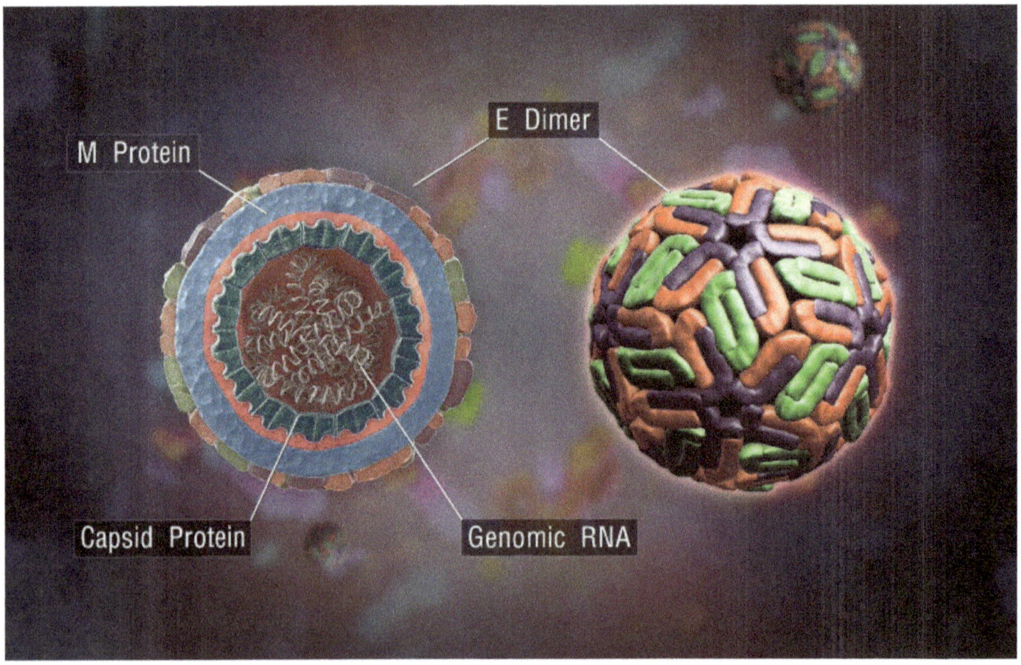

Cross section of a dengue virus showing structural components

E Protein

The DENV E (envelope) protein, found as a dimer on the surface of the mature viral particle, is important in the initial attachment of this particle to the host cell. Each E protein monomer comprises three ectodomains, ED1 to ED3, and a trans-membrane segment. ED2 includes the dimerization interface, two glycosylation sites, and the peptide of fusion with the cellular membrane. ED3 is a continuous polypeptide segment; its fold is compact and immunoglobulin-like. Dengue virus is transmitted by a mosquito known as *Aedes*. Several molecules which interact with the viral E protein (ICAM3-grabbing non-integrin, CD209, Rab 5, GRP 78, and the mannose receptor) have been shown to be important factors mediating attachment and viral entry. The membrane form of Ribosomal protein SA may also be involved in the attachment. Recombinant domains of the E protein are used as well-defined antigens in the serological detection of antibodies directed against dengue virus and as immunogens in vaccine candidates.

prM/M Protein

The DENV prM (membrane) protein, which is important in the formation and maturation of the viral particle, consists of seven antiparallel β-strands stabilized by three disulfide bonds.

The glycoprotein shell of the mature DENV virion consists of 180 copies each of the E protein and M protein. The immature virion starts out with the E and prM proteins forming 90 heterodimers that give a spiky exterior to the viral particle. This immature viral particle buds into the endoplasmic reticulum and eventually travels via the secretory pathway to the Golgi apparatus. As the virion passes through the trans-Golgi Network (TGN) it is exposed to low pH. This acidic environment causes a conformational change in the E protein which disassociates it from the prM protein and causes it to form E homodimers. These homodimers lie flat against the viral surface giving the maturing virion a smooth appearance. During this maturation pr peptide is cleaved from the M peptide by the host protease, furin. The M protein then acts as a transmembrane protein under the E-protein shell of the mature virion. The pr peptide stays associated with the E protein until the viral particle is released into the extracellular environment. This pr peptide acts like a cap, covering the hydrophobic fusion loop of the E protein until the viral particle has exited the cell.

NS3 Protein

The DENV NS3 is a serine protease, as well as an RNA helicase and RTPase/NTPase. The protease domain consists of six β-strands arranged into two β-barrels formed by residues 1–180 of the protein. The catalytic triad (His-51, Asp-75 and Ser-135), is found between these two β-barrels, and activity is dependent on the presence of the NS2B cofactor. This cofactor wraps around the NS3 protease domain and becomes part of the active site. The remaining NS3 residues (180–618), form the three subdomains of the DENV helicase. A six-stranded parallel β-sheet surrounded by four α-helices make up subdomains I and II, and subdomain III is composed of 4 α-helices surrounded by three shorter α-helices and two antiparallel β-strands.

NS5 Protein

The DENV NS5 protein is a 900 residue peptide with a methyltransferase domain at its N-terminal end (residues 1–296) and a RNA-dependent RNA polymerase (RdRp) at its C-terminal end (residues 320–900). The methyltransferase domain consists of an α/β/β sandwich flanked by N-and C-terminal subdomains. The DENV RdRp is similar to other RdRps containing palm, finger, and thumb subdomains and a GDD motif for incorporating nucleotides.

Complexes between the E Protein and Neutralizing Antibodies

Crystal structures of complexes between antibodies and either the ectodomain (sE) of the viral E protein or its domain 3 (ED3) have helped understand the molecular bases of the virus recognition and neutralization. Some of the epitopes are partially or totally innaccessible in the known structure of the mature virion. The corresponding antibodies are therefore assumed to bind to alternate or transitional conformations of the virus at 37 °C.

- The murine antibody E111 neutralizes DENV1. Its Fab and scFv fragments were crystallized in complex with the ED3 domain of DENV1. Its epitope is located around β-strands C and C' of ED3, and the intervening loop.
- The murine antibody 1A1D-2 strongly neutralizes DENV1, DENV2 and DENV3. Its Fab

fragment was crystallized in complex with the ED3 domain of DENV2. Its epitope straddles β-strands A and G of ED3.

- The murine antibody 2H12 cross-reacts with all four DENV serotypes. It neutralizes the corresponding viruses, except DENV2. Its Fab fragment was crystallized in complex with the ED3 domains of DENV1, DENV3 and DENV4. Its epitope is located around the conserved AB loop of ED3.
- The murine antibody 4E11 neutralizes all four DENV serotypes with varying efficacies. Its scFv fragment was crystallized in complex with the ED3 domains of the four DENV serotypes. Its epitope straddles β-strands A and G of ED3 as does the epitope of 1A1D-2. The structures at 2.0 Å resolution have enabled one to analyze the roles of water molecules within the protein interfaces and the roles of somatic hypermutations outside of these interfaces in the interactions and cross-recognitions.
- The chimpanzee antibody 5H2 potently neutralizes DENV4. Its Fab fragment was crystallized in complex with the sE protein of DENV4. Its epitope is included in domain 1 (ED1) of the E protein.
- The human antibodies Ede1-C10, Ede2-A11 and Ede2-B7 potently neutralize all four DENV serotypes. Their Fab or scFv fragments were crystallized in complex with the sE protein of DENV2. The recognition determinants of these antibodies are at a serotype-invariant site in the E dimer interface and include the exposed side-chains of the E fusion loop and the two conserved glycan side-chains.

Mechanism of Infection

1. Dengue virus' (DENV) E envelope protein binds to a cellular receptor. The exact nature of the cellular receptor has not been fully elucidated.
2. DENV undergoes endocytosis. Acidification of the endosome leads to a conformational change of E, exposing a 'fusion peptide' sequence that facilitates fusion of the envelope with the endosomal membrane, releasing the virion capsid into the cytoplasm
3. Uncoating occurs in the cytoplasm
4. Host translational machinery (ribosomes) translates the (+)ssRNA into a single polypeptide
5. Cellular and viral proteinases cleave the polypeptide into 10 proteins (E, M, C and 7 non-structural/enzymatic proteins) while embedded on the ER membrane
6. As soon as functional RNA-dependent RNA polymerase is synthesised RNA replication can commence. Synthesis is asymmetrical, making 10 times more of the positive sense strand than the negative
7. Assembly occurs on intracellular membranes which bud into the ER (forming the envelope from the ER membrane). Subsequent budding from the ER through the Golgi and into

vesicles allows maturation via post-translational modifications eg glycosylation and pH transformational rearrangements

8. Egress occurs via exocytosis

Severe Disease

The reason that some people suffer from more severe forms of dengue, such as dengue hemorrhagic fever, is multifactorial. Different strains of viruses interacting with people with different immune backgrounds lead to a complex interaction. Among the possible causes are cross-serotypic immune response, through a mechanism known as antibody-dependent enhancement, which happens when a person who has been previously infected with dengue gets infected for the second, third or fourth time. The previous antibodies to the old strain of dengue virus now interfere with the immune response to the current strain, leading paradoxically to more virus entry and uptake.

Immune System Interaction

In recent years, many studies have shown that flaviviruses, especially dengue virus has the ability to inhibit the innate immune response during the infection. Indeed, the dengue virus has many nonstructural proteins that allow the inhibition of various mediators of the innate immune system response. These proteins act on two levels :

Inhibition of Interferon Signaling by Blocking Signal Transducer

NS4B it is a small hydrophobic protein located in association with the endoplasmic reticulum. It may block the phosphorylation of STAT 1 after induction by interferons type I alpha, beta. In fact, the activity of Tyk2 kinase decreases with the dengue virus, so STAT 1 phosphorylation decreases too. Therefore, the innate immune system response may be blocked. Thus there is no production of ISG. NS2A and NS4A cofactor may also take part in the STAT 1 inhibition.

NS5 : the presence of this 105 kDa protein results in inactivation of STAT2 (via the signal transduction of the response to interferon) when it is expressed alone. When NS5 is cleaved with NS4B by a protease (NS2B3) it can degrade STAT2. In fact, after the cleavage of NS5 by the protease, there is an E3 ligase association with STAT2, and the E3 ligase targets STAT2 for the degradation.

Inhibition of the Type I Interferon Response

NS2B3-b protease complex is a proteolytic core consisting of the last 40 amino acids of NS2B and the first 180 amino acids of NS3. Cleavage of the NS2B3 precursor activates the protease complex.

This protease complex allows the inhibition of the production of type I interferon by reducing the activity of IFN-beta promoter: studies have shown that NS2B3 protease complex is involved in inhibiting the phosphorylation of IRF3. A recent study shows that the NS2B3 protease complex inhibits (by cleaving) protein MITA which allows the IRF3 activation.

Aedes aegypti D7 Saliva Protein

Dengue virus is transmitted by the mosquito species Aedes aegypti. Aedes aegypti produce saliva that contains over one hundred unique proteins, including the protein family D7. Scientists use to believe that the Aedes aegypti saliva, when being transmitted, actually enhanced the dengue virus in the body. It was said that the mosquito's spit could make the virus spread faster due to the weakened immune response of its host. However, a current study has found that the protein D7 hinders the virus transmission into the host cells.

The immune responses of antibodies, that are trying to fight off the foreign virus, actually increase transmission and make the infection worse. Scientists found levels of protein D7 to be more prevalent in salivary glands of dengue-infected mosquitoes compared to those uninfected. D7 is found in mosquito saliva and was thought to assist the process of blood feeding. Despite the prior assumptions, D7 can modulate the host cell and act against the virus to prevent virus infection. Unfortunately D7 proteins provoke immune responses, which raise anti-D7 antibody levels. These antibodies inhibit the function of D7 proteins, which enhance transmission of dengue virus.

Vaccine Research

Only one vaccine for dengue is currently approved in 3 countries (Brazil, Mexico, Philippines). Several vaccines are under development by private and public researchers. Developing a vaccine against the disease is challenging. With four different serotypes of the dengue virus that can cause the disease, the vaccine must immunize against all four types to be effective. Vaccination against only one serotype could possibly lead to severe dengue hemorrhagic shock (DHS) when infected with another serotype due to antibody-dependent enhancement. When infected with dengue virus, the immune system produces cross-reactive antibodies that provide immunity to that particular serotype. However, these antibodies are incapable of neutralizing other serotypes upon reinfection and actually increase viral replication. When macrophages consume the 'neutralized' virus, the virus is able to replicate within the macrophage, causing disease. These cross-reactive, ineffective antibodies ease access of virus into macrophages, which induces more severe disease (dengue hemorrhagic fever, dengue shock syndrome). A common problem faced in dengue-endemic regions is when mothers become infected with dengue; after giving birth, offspring carry the immunity from their mother and are susceptible to hemorrhagic fever if infected with any of the other three serotypes. One vaccine was in phase III trials in 2012 and planning for vaccine usage and effectiveness surveillance had started.

In 2009 Sanofi-Pasteur started building a new facility in *Neuville-sur-Saône* (fr), a suburb of Lyon (France). This unit produces 4 serotypes vaccine for phase III trials. In September 2014 Sanofi-Pasteur CEO gave early results of the phase III trial efficacy study in Latin America. The efficacy per serotype (ST) varied widely, from 42.3% for ST2, 50.3% for ST1, 74.0% for ST3 and 77.7% for ST4. The full analysis of data from the phase III Latin American-Caribbean study will be reviewed by external experts before being published in a peer-reviewed scientific journal. Primary results has to be presented at the American Society of Tropical Medicine and Hygiene Annual Meeting, held November 2–6, 2014 in New Orleans.

In September 2012, it was announced that one of the vaccines had not done well in clinical trials.

Transmission and Spread of Flavivirus

Virus enters the human body through the mosquito saliva and localizes to various target organs including lymph nodes and liver and starts replication. The virus is then released from target organs and reaches to other lymphatic tissues through blood circulation. Finally virus starts circulating in blood following its release from the lymphatic tissues.

In general the cycle begins with a dengue-infected person. When an uninfected female *Aedes* bites to dengue-infected person during his viremic stage, the virus gets transmitted into the gut of the mosquito. The virus replicates during certain period of time within the mosquito which is called as extrinsic incubation period. The infected mosquito then bites to a susceptible person and transmits the virus, as well as to every other person for its entire life. The virus replicates in the second person and start producing symptoms, this is called as intrinsic incubation period (Inside human body).

Dengue Virus Transmission Cycle between Human and Mosquitoes

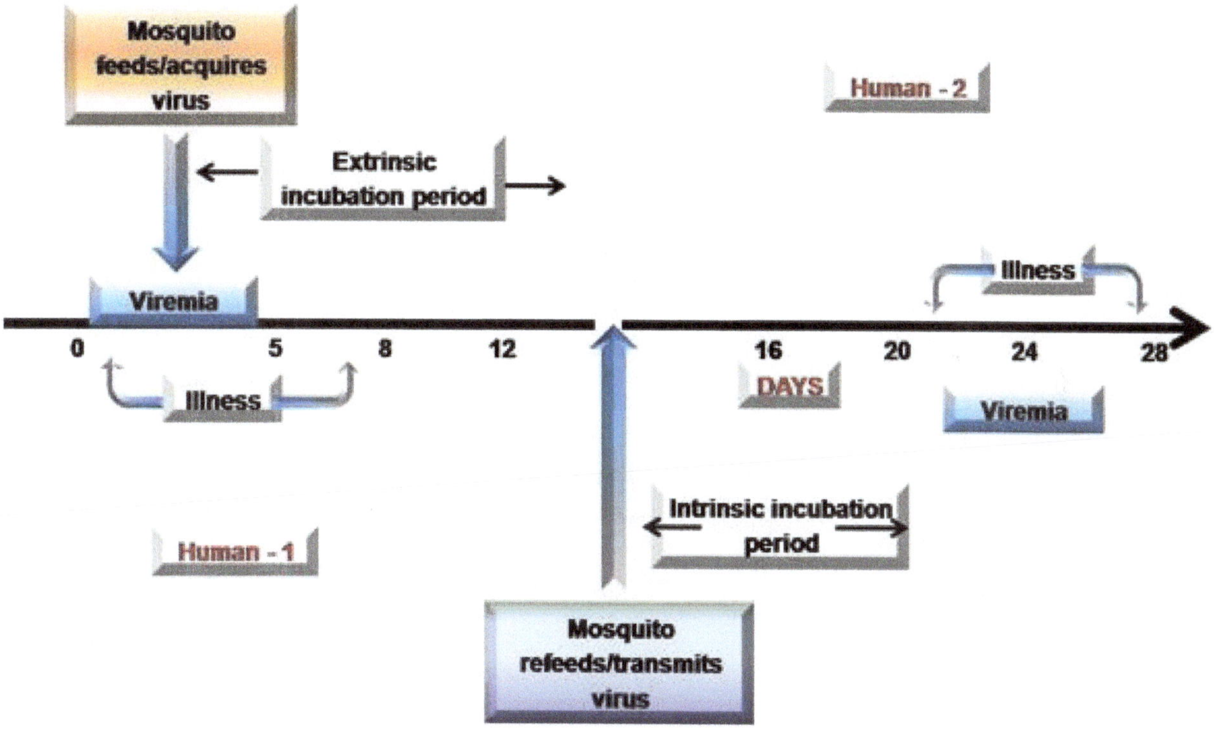

Clinical Features

The dengue virus infection causes fever, headache, joint pain, muscle pain, vomiting, and sometime hemorrhages. Patients may suffer from severe depression after acute form of the disease. Sequential infection with different serotypes of dengue virus put patient into greater risk of developing dengue hemorrhagic fever (DHF) and dengue shock syndrome (DSS). DSS is the severe form of DHF. DSS is characterized by week and rapid pulse, low blood pressure, and altered mental status.

Clinical Feature of Dengue Shock Syndrome

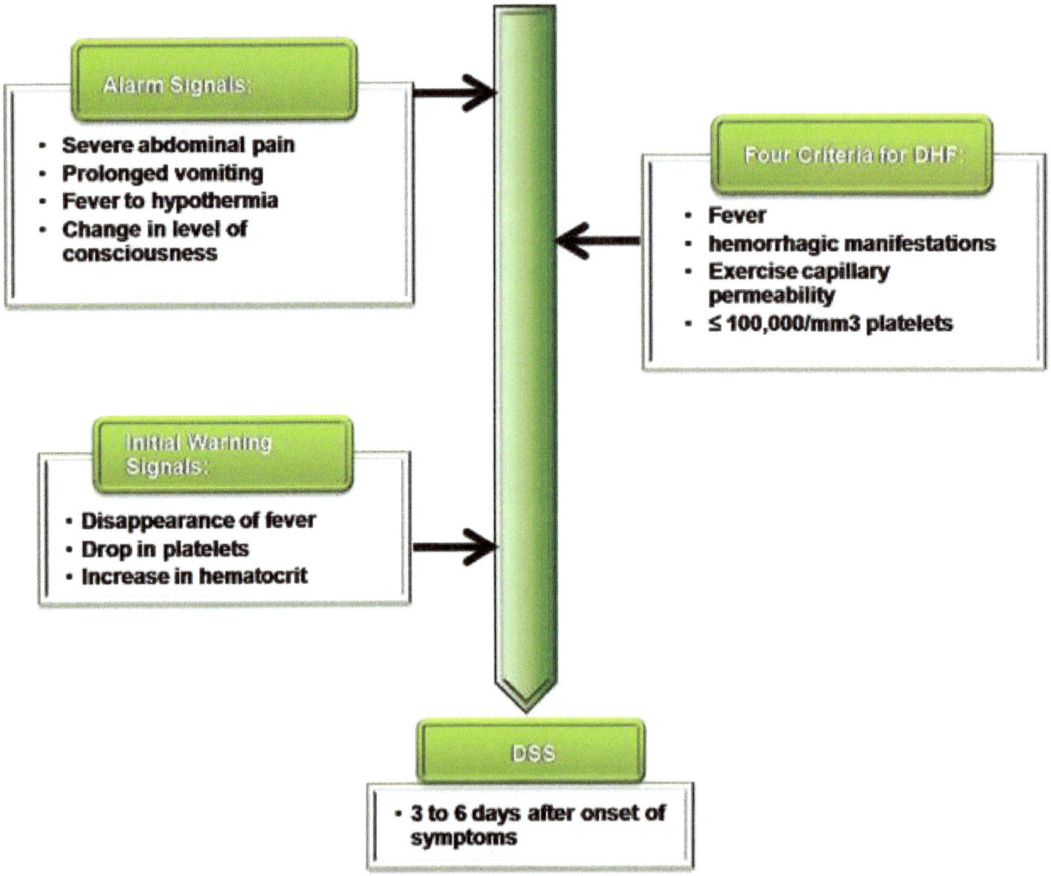

Control

Disease can be controlled by eliminating the female *Aedes* mosquito population. Educating the people to carry out vector control program in their homes and neighborhoods can also help to some extent.

Coronaviruses

Family *Coronaviridae* contains two genera, coronavirus and torovirus . Coronaviruses contain pathogens that mainly infect mammals and birds causing respiratory, gastroenteritis, reproductive and generalized infections. The toroviruses are so far reported in horses and cattle and are associated with diarrhea.

Important Members of Genus Coronavirus

1. Transmissible gastro-enteritis virus

It is a highly contagious disease of swine. Clinical signs of this disease include vomiting, diarrhea, weight loss and dehydration. Disease is more severe in case of young piglets. The virus infects and destroys the enterocytes and intestinal villi resulting in loss of mucosal surface. The measure pathology is restricted to gastro intestinal tract.

2. Feline enteric coronavirus

It is systemic and fatal disease of cats. Clinical signs include fever, anorexia (Loss of appetite), and weight loss. Sometimes ocular and neurological signs are also visible. Virus replicates mostly in monocyte and macrophages.

3. Avian Infectious bronchitis virus

It is the respiratory disease of poultry characterized by cough, gasping, respiratory rales and dysponea. Clinical signs include visceral gout, nephritis and cheesy exudate in the trachea. Birds may develop secondary bacterial infection.

4. Mouse hepatitis virus

Mouse hepatitis virus is a causative agent of enteric infection in infant mice. It is a highly contagious virus of mice and mortality may approach to 100% in virulent outbreak cases. Symptoms include dehydration and rapid weight loss in infants. Lesions of the disease are visible mainly in small intestine and proximal part of large intestine.

Virion Characteristics

Schematic representation of mouse hepatitis virus:

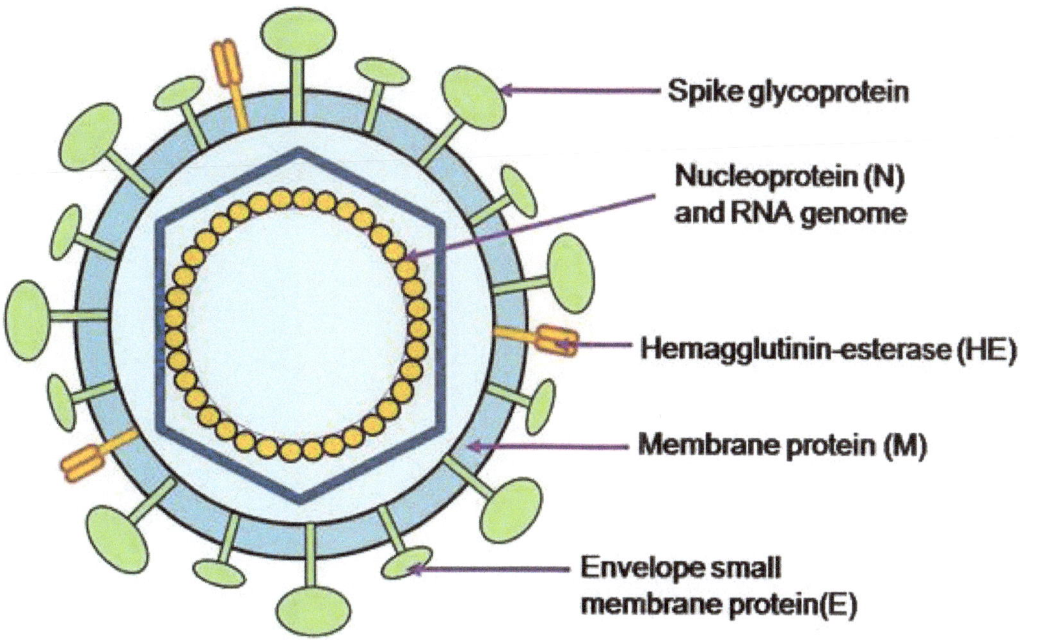

Coronaviruses are enveloped, approximately 80-250 nm in diameter, and are pleomorphic in shape. The viruses contain large club shaped spike protrusion from the surface of the icosahedral internal core. The core contains helical nucleocapsids which is linear single stranded RNA of 25-31 kb in size. The 5' end of the genome is capped and 3' of the genome is polyadenylated similar to that of picornaviruses. The major structural protein of the coronavirus includes nucleoprotein (N), spike protein (S), membrane protein (M), envelope protein (E), and hemagglutinin esterase protein (HE). The S protein is a major antigenic determinant of the virus and contains the conformational epitopes towards their N terminal. Host immune responses are directed mainly against N and S proteins which are the main targets for the vaccine designing strategy.

Virus Replication

Coronaviruses are reported to utilize many receptors for their entry inside the cells. Severe acute respiratory syndrome (SARS) coronavirus uses angiotensin-converting enzyme 2 (ACE-2), whereas mice hepatitis virus uses carcinoembryonic antigen related cell adhesion molecule 1 (CEACAM-1) while others use sialic acid as a receptor.

Schematic representation of replication scheme in mouse hepatitis virus:

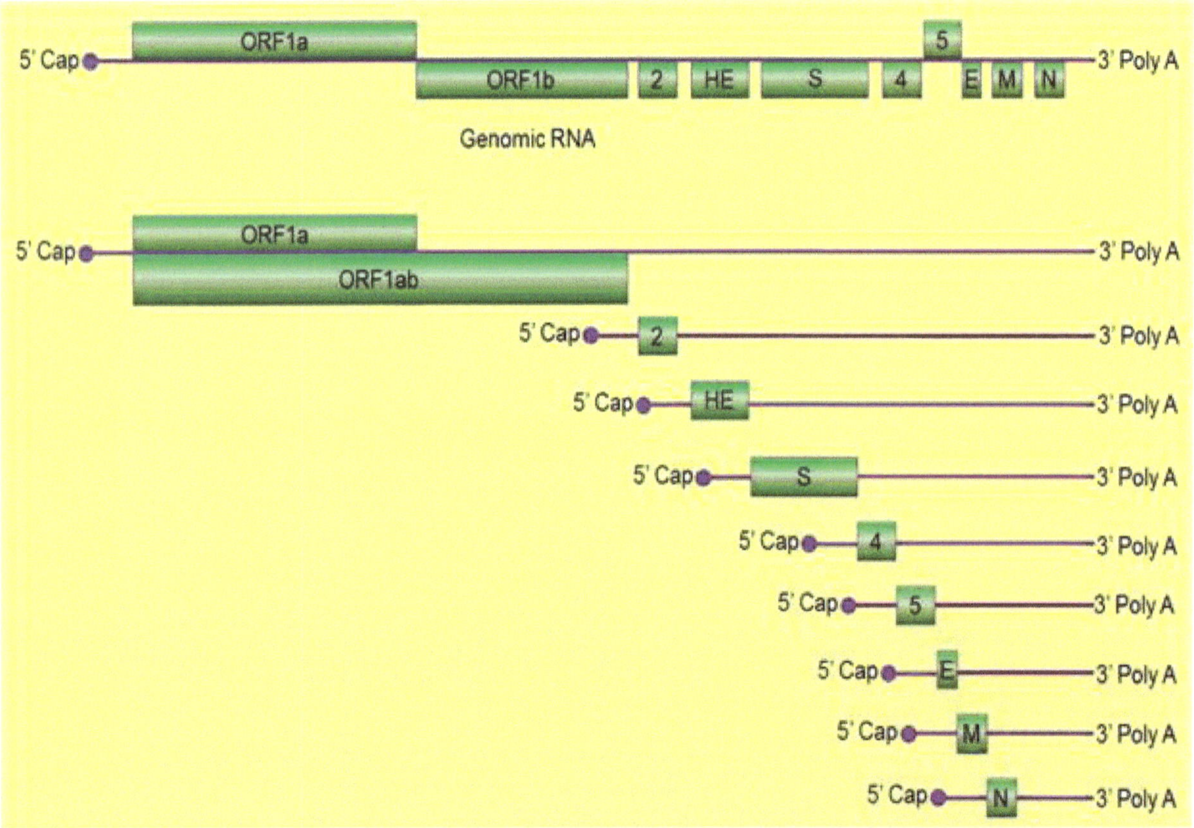

Viral RNA serves as an mRNA for the synthesis of RNA dependent RNA polymerase (RDRP). Two large open reading frames are translated as a single polyprotein which are later cleaved into separate viral protein. The RDRP then forms the full length complementary genome which serves as the template for the formation of subgenomic RNA. All the mRNAs are then translated into various viral proteins. Many of the viral proteins undergo posttranslational modifications in endoplasmic reticulum and golgi body and therefore the progeny virions buds out from these location and not from cell membrane.

Overall view of coronavirus infectious cycle:

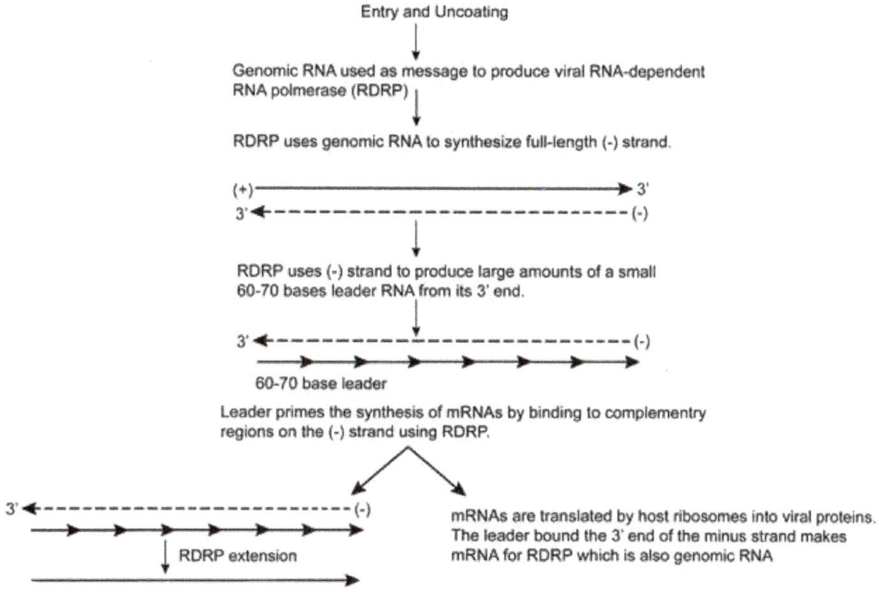

SARS Pathogenesis

Severe acute respiratory syndrome (SARS) is a highly infectious viral disease caused by members of family *Coronaviridae* . The first case was recognized in 2003 with signs of atypical pneumonia. SARS i s caused by a novel coronavirus (CoV) which can infect a wide range of animal species. Many SARS-CoVs has been found in bats suggesting it to be the reservoir of the virus. The first outbreak of SARS was notified in Southern China where the bats and humans lived in the close proximity with each other. It mainly affects the adult human beings. Transmission is through direct contact with infected patient or through infected body fluids. Incubation period varies from 5 days to 2 weeks. Symptoms include cough, shortness of breath, and difficulty in breathing with mild fever.

Virion Properties

The viral genome of SARS-CoV contains ssRNA of 29.7 kb in size, one of the largest among RNA viruses group. The virion is spherical in shape around 80-100 nm in diameter. Various projections from the surface of the virion protrude to form a crown like structure (hence named coronavirus). 5' end of the virus encodes for an open reading frame which is translated into a large polyprotein that is cleaved by viral proteases to form many non-structural proteins. The non structural proteins include RNA dependent RNA polymerase, ATPase and helicase. The non structural proteins helps in replicating viral genome as well as viral transcripts and subgenomic mRNA's which are used to synthesize viral proteins. The viral membrane protein includes spike protein, and a membrane protein which completes its maturation in endoplasmic reticulum and golgi apparatus. The spike (S) protein present on the surface of the virion is the major antigenic determinant of the virus. Spike protein is responsible for tissue tropism as well as gradient of pathogenesis of SARS-CoV. The virus infects the cells of the respiratory epithelium by binding to its receptor, angiotensin-converting enzyme 2 (ACE-2).

Pathophysiology

Respiratory and gastrointestinal tract are the only organs reported to support SARS-CoV replication. The virus invades the respiratory epithelium and cause lytic effect on the infected cells. The effects are more pronounced in the pulmonary tissues where wide spread alveolar damages are visible. Although the morbidity and mortality is higher among old age group (65 years), intensive care treatment and mechanical ventilation can improve the life expectancy of the patients.

Clinical Picture of SARS:

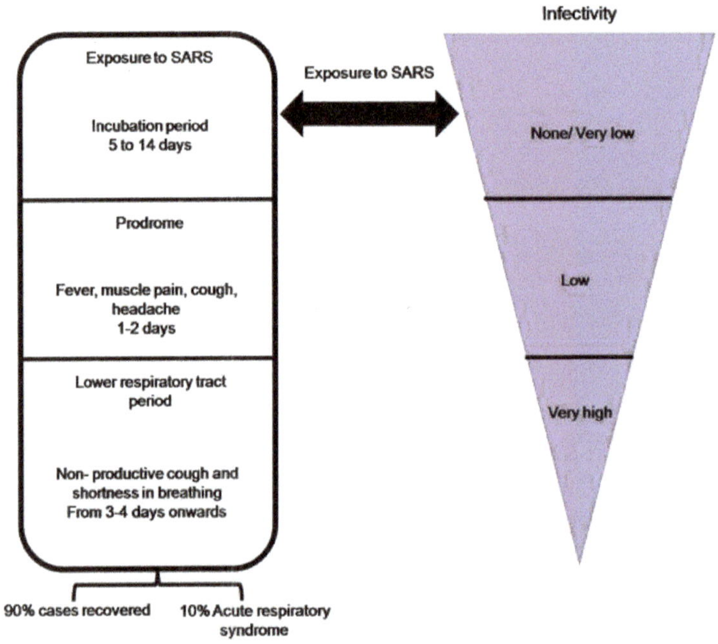

Diagnosis

Diagnosis is based on

I. PCR for identifying SARS-CoV genome sequence.

II. Enzyme-linked immunosorbent assay (ELISA) from serum samples collected from patient.

III. Immunofluorescence test (IFT) from serum samples collected from patient.

IV. Virus isolation from infected tissue samples using cell culture techniques.

Treatment

Normal treatment is mostly symptomatic and includes followings:-

I. Antibiotics to treat secondary bacterial infections

II. Antiviral drugs

III. Corticosteroids to reduce pathology of lungs

IV. Ventilators and artificial breathing

Prevention and Control

a) Reduce the contact with SARS infected individuals

b) Hand washing and personal hygiene

c) Covering mouth and nose while sneezing to avoid droplet infection

d) Avoid sharing food and drinks from an infected person

e) Disinfection of the working space and surroundings.

References

- Conway MJ (2015). "Identification of a flavivirus sequence in a marine arthropod". PLOS ONE. 10 (12): e0146037. doi:10.1371/journal.pone.0146037. PMC 4699914 . PMID 26717191
- Patton JT (editor). (2008). Segmented Double-stranded RNA Viruses: Structure and Molecular Biology. Caister Academic Press. ISBN 978-1-904455-21-9
- Henderson BR, Saeedi BJ, Campagnola G, Geiss BJ (2011). Jeang, K.T, ed. "Analysis of RNA binding by the Dengue virus NS5 RNA capping enzyme". PLoS ONE. 6 (10): e25795. Bibcode:2011PLoSO...625795H. doi:10.1371/journal.pone.0025795
- "Virus Taxonomy Portal." (Website.) Viral Bioinformatics Resource Center & Viral Bioinformatics - Canada. Retrieved on 2007-09-27
- Roy P (2008). "Molecular Dissection of Bluetongue Virus". Animal Viruses: Molecular Biology. Caister Academic. ISBN 978-1-904455-22-6
- Sun, G.; Larsen, C.; Baumgarth, N.; Klem, E; Scheuermann, R. (26 January 2017). "Comprehensive Annotation of Mature Peptides and Genotypes for Zika Virus". PLoS ONE. 12: e0170462. PMC 5268401 . PMID 28125631. doi:10.1371/journal.pone.0170462. Retrieved 20 February 2017
- West, Christy (2010-02-08). "Different West Nile Virus Genetic Lineage Evolving?". The Horse. Retrieved 2010-02-10. From statements by Orsolya Kutasi, DVM, of the Szent Istvan University, Hungary at the 2009 American Association of Equine Practitioners Convention, December 5–9, 2009
- Le Sommer C, Barrows NJ, Bradrick SS, Pearson JL, Garcia-Blanco MA (2012). Michael, Scott F, ed. "G protein-coupled receptor kinase 2 promotes flaviviridae entry and replication". PLoS Negl Trop Dis. 6 (9): e1820. doi:10.1371/journal.pntd.0001820
- Martinez MA, et al. (2012). "Quasispecies Dynamics of RNA Viruses". In Witzany, G. Viruses: Essential Agents of Life. Springer. pp. 21–42. ISBN 978-94-007-4898-9
- Bidet, Katell; Garcia-Blanco, Mariano A. (2014-09-01). "Flaviviral RNAs: weapons and targets in the war between virus and host". Biochemical Journal. 462 (2): 215–230. ISSN 0264-6021. PMID 25102029. doi:10.1042/BJ20140456
- Roy P (2008). "Structure and Function of Bluetongue Virus and its Proteins". Segmented Double-stranded RNA Viruses: Structure and Molecular Biology. Caister Academic. ISBN 978-1-904455-21-9
- Ghosh Roy S, Sadigh B, Datan E, Lockshin RA, Zakeri Z (2014). "Regulation of cell survival and death during Flavivirus infections". World Journal of Biological Chemistry. 5 (2): 93–105. PMC 4050121 . PMID 24921001. doi:10.4331/wjbc.v5.i2.93
- "A Second Time Dengue Infection Can be Life-Threatening Says Research". NDTV. United Kingdom. September 18, 2015. Retrieved 2016-05-07

- Shi, P-Y (editor) (2012). Molecular Virology and Control of Flaviviruses. Caister Academic Press. ISBN 978-1-904455-92-9
- McLean JE, Wudzinka A, Datan E, Quaglino D, Zakeri Z (2011). "Flavivirus NS4A-induced autophagy protects cells against death and enhances virus replication". Journal of Biological Chemistry. 286 (25): 22147–22159. PMC 3121359 . PMID 21511946. doi:10.1074/jbc.M110.192500
- Halstead, SB (1988). "Pathogenesis of dengue: challenges to molecular biology.". Science. 239 (4839): 476–481. Bibcode:1988Sci...239..476H. PMID 3277268. doi:10.1126/science.3277268. Retrieved 23 February 2013
- WHO (2009). Dengue Guidelines for Diagnosis, Treatment, Prevention and Control (PDF). World Health Organization. ISBN 92-4-154787-1
- Miller JL, de Wet BJ, deWet BJ, et al. (February 2008). "The mannose receptor mediates dengue virus infection of macrophages". PLoS Pathog. 4 (2): e17. PMC 2233670 . PMID 18266465. doi:10.1371/journal.ppat.0040017
- Hanley, K.A.; Weaver, S.C., eds. (2010). Frontiers in Dengue Virus Research. Caister Academic. ISBN 978-1-904455-50-9
- Conway MJ (2015). "Identification of a flavivirus sequence in a marine arthropod". PLOS ONE. 10 (12): e0146037. PMC 4699914 . PMID 26717191. doi:10.1371/journal.pone.0146037

4

Negative Sense RNA Viruses

Negative sense RNA viruses are a part of the Mononegavirales order. Owing to its negative polarisation, these viruses need to be converted into positive RNA and therefore is not infectious by themselves. Its genome has features like linear non-segmented negative sense RNA genome, the leader RNA is neither capped nor polyadenylated, and its helical nucleocapsid contains the RNA dependent RNA polymerase among other features. The major categories of negative sense RNA viruses are dealt with great details in the section.

Negative-sense Single-stranded RNA Virus

A negative-sense single-stranded RNA virus is a virus that uses negative sense, single-stranded RNA as its genetic material. Single stranded RNA viruses are classified as positive or negative depending on the sense or polarity of the RNA. The negative viral RNA is complementary to the mRNA and must be converted to a positive RNA by RNA polymerase before translation. Therefore, the purified RNA of a negative sense virus is not infectious by itself, as it needs to be converted to a positive sense RNA for replication. These viruses belong to Group V on the Baltimore classification.

In addition, negative-sense single-stranded RNA viruses have complex genomic sequences, cell cycles, and replication habits which utilize various complexes of proteins in order to arrange itself in specific conformations and carry out necessary processes for proper survival and reproduction of its genomic sequence. The complexity of negative-sense single-stranded RNA viruses carries into its ability to suppress the innate immune response of the cells it infects and the construction of a capsid which is unique to the varying classifications of negative-sense single-stranded RNA viruses.

Replication

Negative sense ssRNA viruses need RNA polymerase to form a positive sense RNA. The positive-sense RNA acts as a viral mRNA, which is translated into proteins for the production of new virion materials. With the newly formed virions, more negative sense RNA molecules are produced.

In more details, replication of the virus consists of the following steps:

1. A virion enters the host cell and releases its negative RNA into the cytoplasm.
2. The virus uses its own RNA replicase, also known as RNA-dependent RNA polymerase (RdRp), to form positive RNA template strands through complementary base pairing.
3. The positive RNA acts as mRNA, which is translated into structural capsomere proteins and viral RdRp by the host's ribosomes.

4. A replicative complex is formed with RdRp: The positive strands can either function as mRNA to produce more proteins or as template to make more negative RNA strands.
5. New viral capsids are assembled with the capsomere proteins. The negative RNA strands combine with capsids and viral RdRp to form new negative RNA virions.
6. After assembly and maturation of nucleocapsid, the new virions exit the cell by budding or lysing through cell membrane to further infect other cells.

The genome size of a negative RNA virus is between 10kb to 30kb. Two genome subgroups can be distinguished, nonsegmented and segmented, and are described as such:

- In viruses with nonsegmented genomes, the first step of replication is transcription of the negative strand by RdRp to form various monocistronic mRNA that code for individual viral proteins. A positive strand copy is formed to serve as template for the production of the negative genome. This replication takes place in the cytoplasm.
- In viruses with segmented genomes, replication occurs in the nucleus and the RdRp produces one monocistronic mRNA strand from each genome segment. The principal difference between the two types is the location of replication.

Taxonomy

One order and twelve families are currently recognised in this group. Several of these families contain extremely similar genetic structures and are classified under the order *Mononegavirales*. A number of unassigned species and genera are yet to be classified.

- Order *Mononegavirales*
    - Family *Bornaviridae* — Borna disease virus
    - Family *Filoviridae* — includes Ebola virus, Marburg virus
    - Family *Mymonaviridae*
    - Family *Nyamiviridae*
    - Family *Paramyxoviridae* — includes Measles virus, Mumps virus, Nipah virus, Hendra virus, and NDV
    - Family *Pneumoviridae* — includes RSV and Metapneumovirus
    - Family *Rhabdoviridae* — includes Rabies virus
    - Family *Sunviridae*
    - Genus Anphevirus
    - Genus Arlivirus
    - Genus Chengtivirus

  - Genus Crustavirus
  - Genus Wastrivirus
- Unassigned families:
  - Family *Arenaviridae* — includes Lassa virus
  - Family *Bunyaviridae* — includes Hantavirus, Crimean-Congo hemorrhagic fever
  - Family *Ophioviridae*
  - Family *Orthomyxoviridae* — includes Influenza viruses
- Unassigned genera:
  - Genus *Deltavirus* — includes Hepatitis D virus
  - Genus *Emaravirus*
  - Genus *Tenuivirus*
- Unassigned species:
  - Taastrup virus

Host Range

Viruses of the families *Arenaviridae*, *Orthomyxoviridae*, *Paramyxoviridae*, and *Pneumoviridae* are able to infect vertebrates. Viruses of the families *Bunyaviridae* and *Rhabdoviridae* are able to infect vertebrates, arthropods, and plants. Viruses of the family *Tenuivirus* only infect plants. A few viruses known to infect humans include Marburg virus, Ebola, measles, mumps, rabies, and influenza.

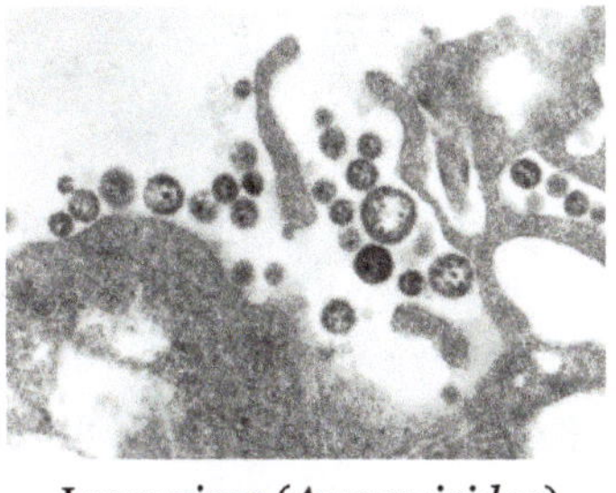

Lassa virus (*Arenaviridae*)

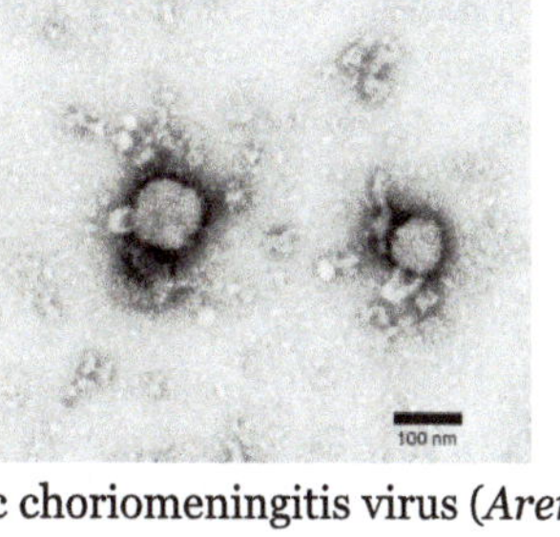

Lymphocytic choriomeningitis virus (*Arenaviridae*)

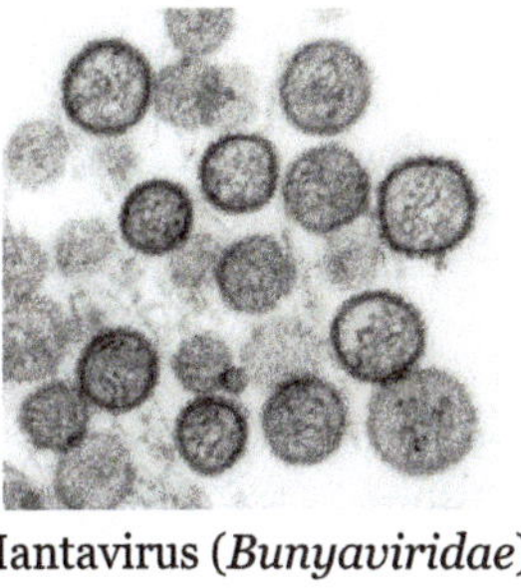

Hantavirus (*Bunyaviridae*)

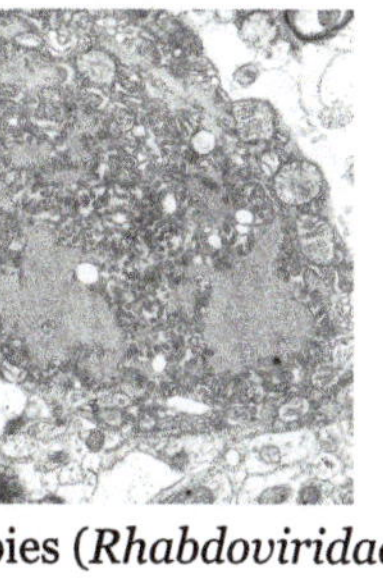

Rabies (*Rhabdoviridae*)

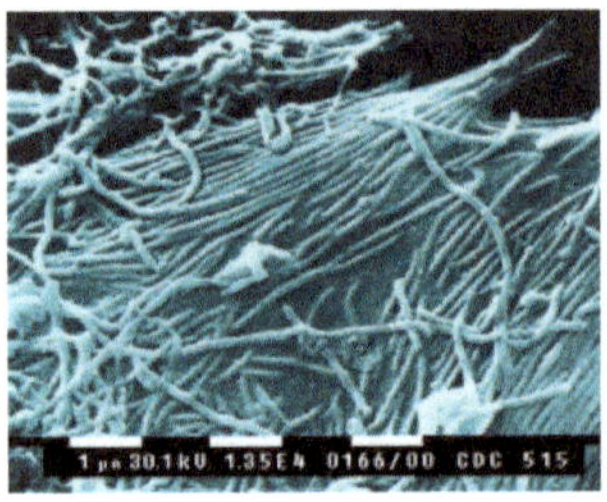

Ebola virus (*Filoviridae*)

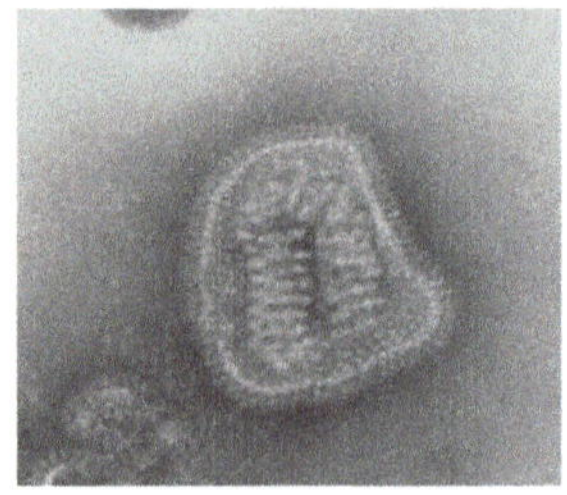

Influenza (*Orthomyxoviridae*)

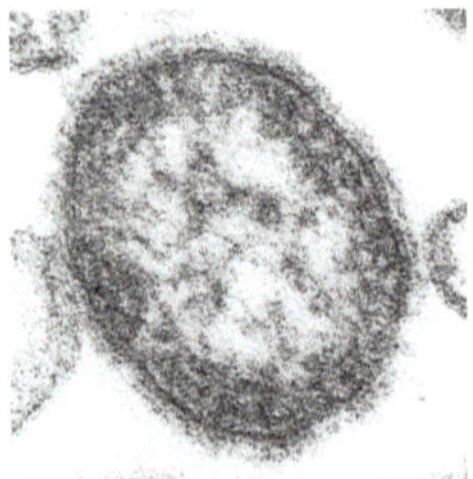

Measles (*Paramyxoviridae*)

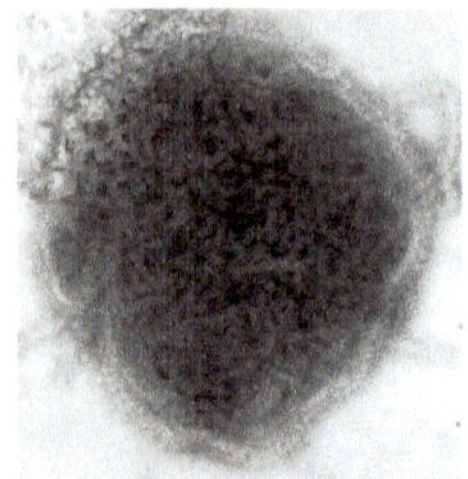

Mumps virus (*Paramyxoviridae*)

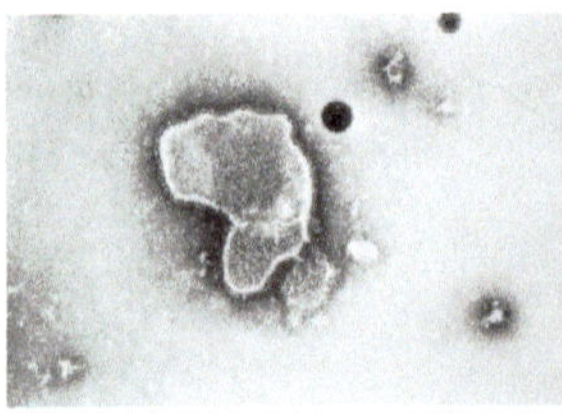

Human respiratory syncytial virus (*Paramyxoviridae*)

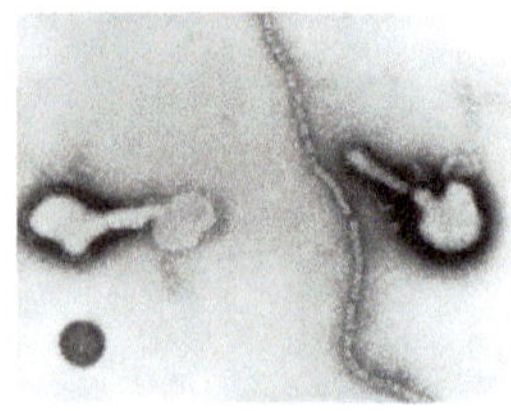

Parainfluenza (*Paramyxoviridae*)

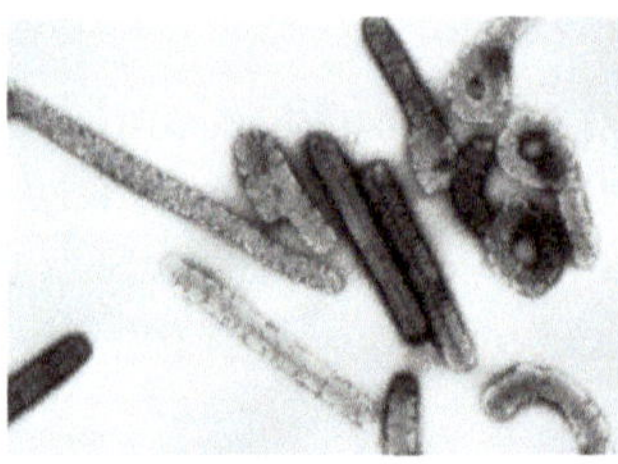

Marburg Virus (*Filoviridae*)

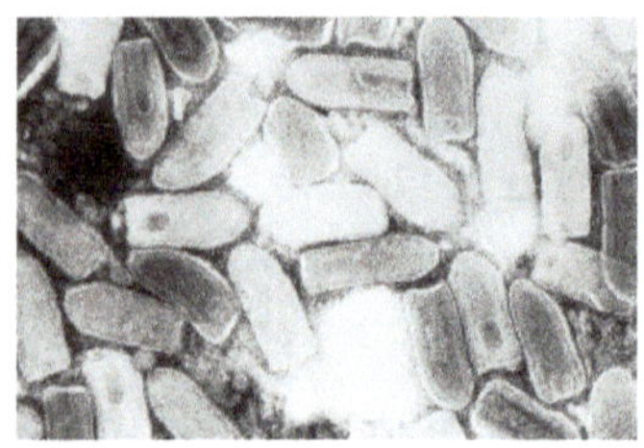

Vesicular stomatitis virus (*Rhabdoviridae*)

Genome

The genome for negative-stranded RNA virus (NSV) consists of one to several single-stranded RNAs, which are assembled into complexes. These complexes function as templates for the transcription and replication of the NSV genome by RNA polymerase. The NSV genome has been observed as being either segmented and nonsegmented. The segmented genome can be divided up into anywhere between 2-8 RNA molecules. A distinctive feature of the NSV genome is its highly structured organization. Multiple monomers of either the ribonucleoprotein (RNP) complexes or nucleocapsids associate with the RNA in order to create these organized complexes. Nonsegmented RNA follows a sequential pattern of gene expression. With segmented RNA, each individual segment is contained in a distinct RNP complex. These complexes function independently for transcription and replication related processes.

NSV ribonucleoproteins (RNPs) generally adopt a helical conformation. For nonsegmented NSVs, these conformations are typically linear in fashion, and are relatively rigid. Segmented NSVs are present in a more flexible, circular conformation. This results from non-covalent RNA-RNA interactions between the 5'- and 3'-termini of the RNA segments. Regarding the helical character of NSV RNPs, the main determinant of this conformation is the NP/N protein. The structure of virion full-length RNPs are typically presented as a left-handed, double-helical arrangement of two NP strands that are opposite in polarity. Recombinant RNPs have a different structure, generally consisting of a right-handed helix.

Catalytic activity related to RNA synthesis is associated with a very large protein, termed the L protein. This is a multi-enzymatic polypeptide that is responsible for multiple tasks. These activities span multiple domains, including mRNA synthesis/modification and the formation of ribonucleoproteins. Through phylogenetic analysis, it was determined that NSV RNA polymerases share a common ancestor with other RNA polymerases from various origins. This is further supported by various L proteins displaying highly conserved sequence blocks that are in series, yet separated by variable regions.

When discussing the NSV RNP, the NP/N protein is considered to be the most abundant element of the ribonucleoprotein. These elements are what provide the basis for the RNPs helical conformation. NP/N proteins are also vital for the transcription and replication of full-length RNA templates. The proper function of NP/N proteins rely on their capacity to oligomerize. This defines the ability of NP/N proteins to complex together to form larger structural elements. The oligomerization mechanism is unique between the nonsegmented NSV genome (nsNSV) and segmented NSV genome (sNSV). For nsNSVs, interactions occur between the protein monomers, while the stabilization of this complex occurs through inter-monomeric contact involving N-terminal and/or C-terminal protein extensions. This contributes to the viral RNP helical structure. Oligomerization for sNSVs has been shown to be similar to this for some Bunyaviridae, but it may also take place much more flexibly, defined by their ability to form much smaller oligomers. These include, but are not limited to, dimers, trimers, tetramers, long helices, and more complex structures.

NSV Life Cycle and Replication

Negative-strand RNA viruses (NSV) can be classified into 7 distinct families. The families consisting of nonsegmented genomes include Rhabdo-, Paramyxo-, Filo- and Borna-. Orthomyxo-, Bunya-, Arenaviridae- contain genomes which consist of six to eight, three or two negative-sense RNA segments, respectively. Many highly prevalent human pathogens such as the respiratory syncytial virus (RSV), parainfluenza viruses, influenza viruses, Ebola virus, Marburg virus are included within the NSV. The life cycle of NSV has a number of steps. The virus first infects the host cell by binding to the a host cell receptor through a viral surface glycoprotein. The fusion of the glycoprotein viral membrane with the plasma membrane of the host cell in an acidic environment allows for the release of viral ribonucleoprotein (RNP) complexes into the cytoplasm. Most NSV replicate in the cytoplasm of infected cells. Newly synthesized RNP complexes are assembled with viral structural proteins at the plasma membrane or at membranes of the Golgi apparatus. This is all followed by the release of the newly synthesized viruses.

In regards to the replication and transcription of non-segmented NSV, the genes of these NSV are made up of three regulatory regions: a gene end signal, an intergenic region, and a gene start signal. One example of gene end signals are in a specific virus called the vesicular stomatitis virus (VSV) contains gene end signals that are highly conservative. The intergenic region is highly variable and consists of conserved dinucleotide, trinucleotide, or regions of up to 143 nucleotides. The various lengths of the intergenic regions correlate with transcriptional attenuation, however diverse intergenic regions do not alter the gene expression. The gene start signals are highly specific as the first three nucleotides are critical for the gene expression.

Molecular Mechanisms of Innate Antiviral Immune Inhibition

A cell's innate immune response is the first line of defense for finding viral infections. The innate immune response triggers the production of type I interferons (IFN) and pro-inflammatory cytokines. However, non-segmented negative sense RNA viruses (NNSV) have developed multifunctional proteins which are capable of diffusing this innate response pathway in order to avoid antiviral functions within the cell. This mainly occurs through the NNSV's proteins interacting with other cellular proteins which are involved in the type I interferon signaling pathway. Due to the expansive nature of the immune pathway, NNSVs have a variety of areas throughout the signaling cascade which can be targeted for the interruption of the production of type I interferons or pro-inflammatory cytokines. Such areas include inhibiting the induction of the IFN pathway or inhibiting the response from the IFN signaling cascade.

NNSV-Mediated Inhibition of IFN Inhibition

Several areas of the induction pathway, as outlined in the image provided, are targeted by NNSVs in order to avoid detection within the cell or even inhibit that specific area of the signaling cascade. One such example of NNSVs avoiding cellular detection arises with mutations which sequester the pathogen-associated molecular pathways. By creating a mutation within the nucleotide sequence, specifically ones important for binding double stranded DNA or other proteins, the virus is able to go undetected by cell, avoid activation of the cellular antiviral response, and evade the immune response.

NNSVs can also bind to cellular receptors throughout the pro-inflammatory cytokine pathway in order to inhibit the immune response. By carrying accessory proteins which directly bind to pattern recognition receptors, the virus is able to use its accessory proteins to induce conformational changes throughout other immune response proteins and inhibit cellular responses. Generally, the pattern recognition receptors are responsible for detecting infection associated molecules commonly associated with viruses, but the accessory proteins that some viruses may carry reconfigure the protein in order to inhibit its function and block the rest of the signaling cascade which would originally produce an immune response.

Other areas of inhibition of induction apply similar concepts of binding to cellular proteins and inhibiting their function throughout the immune cascade. These include binding to proteins involved in dephosphorylation pathways and blocking DNA binding transcription factors. In each case, the accessory proteins coded for in the nucleotide sequence inhibit a critical function of other innate cellular proteins, disrupting the signaling cascade for producing type I interferons and pro-inflammatory cytokines.

NNSV-Mediated Inhibition of IFN Response

Another way that NSVs avoid the immune response of the host is to encode for proteins which target the JAK/STAT pathway or the nuclear transports mechanisms for transcription factors.

Each of these are a portion of the IFN pathway described previously in the immune system's innate response to viral infection.

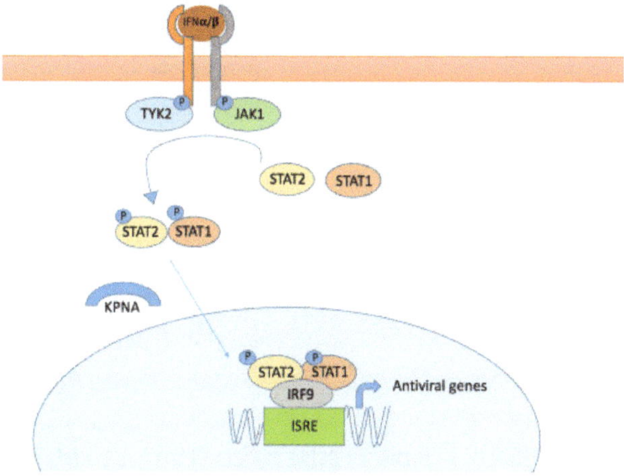

JAK/STAT pathway

For the JAK/STAT pathway depicted in the image, a critical reaction for inducing the pathway is the proper phosphorylation of the TYK2 and JAK1 proteins. Upon phosphorylation, the rest of the STAT pathway will begin and lead to the production of antiviral genes. NNSVs have the capability to synthesize proteins which target the phosphorylation step of the pathway. By preventing the phosphorylation of TYK2 and JAK1, they stop the IFNα/β pathway and bring the antiviral immune response to a halt. Similarly, viruses can also synthesize proteins which will prevent the phosphorylation of STAT1 a little further along the signaling cascade. This process has the same result in halting the IFNα/β pathway, just like preventing phosphorylation of TYK2 and JAK1.

Further down the IFNα/β pathway, STAT1 and STAT2 are transported across the nuclear membrane, as depicted in the image. They accordingly bind to the DNA sequence and behave as transcription factors- affecting the level of gene production throughout the cell. In normal cellular function, this pathway will behave normally in response to viral infection, leading to the production of antiviral genes and the induction of an immune response. However, NNSVs have developed the capability to generate complexes which target the protein responsible for translocating STAT1 across the nuclear membrane. By binding to and inhibiting this function, STAT1 is never able to bind to the DNA and properly regulate the production of antiviral genes. This pathway of inhibiting the IFN response helps the NNSV go undetected within the cell and avoid certain immune response pathways.

Ubiquitination

Another important process utilized throughout many cellular processes is ubiquitination- which is outlined in the image below. This is utilized in many cells in order to locate and identify viruses, as well as, restrict the viral infection throughout. NNSVs however have developed a pathway to synthesize proteins which target the ubiquitin pathway along many of the signaling cascades descriptive of the IFN response. More specifically, the NNSVs are capable of reprogramming the host cell's ubiquitination pathway in such a way that leads to the degradation of host cell mechanism that would otherwise silence a viral infection.

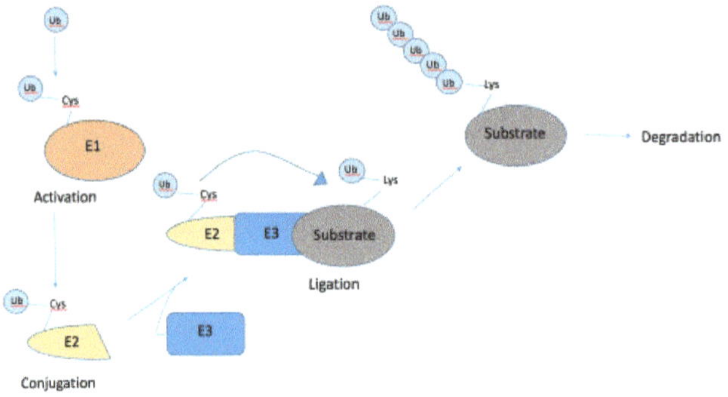

Ubiquitination Pathway

Common Mechanism for RNA Encapsidation by NSVs

Like all viruses, negative-sense RNA viruses (NSVs) contain a protein capsid that encapsulates the genomic material. The nucleocapsid of NSVs is assembled with a single nucleocapsid protein, and the viral RNA. Each NSV nucleocapsid is packaged inside a lipid envelope, but the appearance of the nucleocapsid differs from virus to virus. For example, in rhabdoviruses, the nucleocapsid assumes a bullet shape, while in paramyxoviruses, the nucleocapsid is filamentous or herringbone-like. However, when the nucleocapsids are released from the virion, they all appear like a coil.

So far, the atomic structures of nucleocapsid-like-particles (NLP) have been elucidated for three NSV families: Rhabdoviridae, Paramyxoviridae, and Bunyaviridae. One important element of NSVs is that the capsid protein (N) is first synthesized as a monomeric protein (No). No is a capsid protein that assembles a capsid to accommodate any RNA sequence. No is kept monomeric in different ways depending on which family of NSVs a virus falls into. For instance, rhabdoviruses keep the No monomeric by forming a complex with the phosphoprotein (P). It has been found that both the N- and C terminal regions of P bind to the capsid protein. Essentially, the P binding occupies the site required for No oligomerization. Other viruses, such as bunyaviruses, simply sequester the N terminus by N itself. Nonetheless, it is essential for N to be monomeric in order for the NSV to be competent in encapsidating viral RNA.

In terms of the structure of the nucleocapsid, the N protein will eventually oligomerize to encapsidate the single-stranded RNA. In some NSVs the N subunits are oriented in a parallel orientation, and the ssRNA is sequestered in the center. In most NSVs, the nucleocapsid appears to be a random coil, and the symmetry is linear. Because the N subunits are oriented in a parallel fashion, the cross-molecular interactions among the subunits stabilize the nucleocapsid and are critical for capsid formation. The linear interactions along the encapsidated single-stranded RNA are actually a unique feature to NSVs. X-ray crystal structures of N proteins and EM micrographs of RNP complexes from a number of Bunyaviridae (a Family of segmented NSVs that includes the Bunyamwera virus and the Schmallenberg virus) show that in these viruses the nucleocapsid adopts a helical arrangement where the N proteins form a head-to-tail chain by linking to each other via a flexible N-terminal arm. The resulting chain forms tight coils with four N proteins per turn that holds the circular ssRNA of these viruses on the inside of the coils and which extend to form large circular filaments.

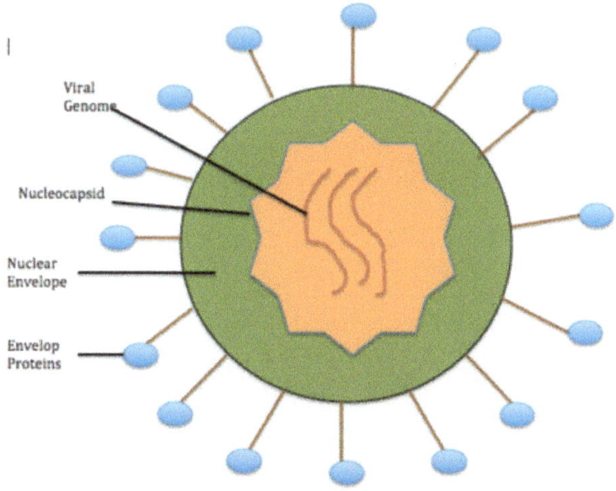

Structure of a Nucleocapsid

The process of encapsidation is a concomitant process with viral replication, likely at the site of the viral RNA replication process. Another unique feature of NSVs is that the sequestered bases are stacked to form a motif similar to half of the A-form double helix of RNA. The stacking arrangement allows for maximized packaging of the RNA in the capsid. The base stacking may actually contribute to the overall stability of the nucleocapsid, and is dependent on the RNA sequence (i.e. poly(rA) is the most stable). An interesting finding is that specific sequences in the sequestered RNA genome may regulate viral functions. For translational termination, there is a U7 track at the end of each coding region, causing this to be the least stable region. This therefore promotes dissociation of the transcription complex. A unique aspect of NSVs is the conserved (3H+5H) motif, which has been identified to constitute the RNA cavity through secondary structure elements of the N protein.

Lastly, a main distinction between NSVs from other viruses is that the nucleocapsid actually serves as the template for viral RNA synthesis. During synthesis, the nucleocapsid undergoes a conformational change to release the RNA template. After transcription is complete, the RNA is repositioned in the cavity, and the nucleocapsid is restored.

Features of RNA Genome

I. Linear non-segmented negative sense RNA genome

II. Organization of genome- 3'-Leader-Virion core- Surface proteins-Polymerase-Trailer 5'.

III. Helical nucleocapsid contains the RNA dependent RNA polymerase.

IV. The leader RNA is neither capped nor polyadenylated and is not functional as mRNA.

V. Replication occurs when the polymerase complex ignores the transcription stop signals at the 3' end of each gene and a full-length positive-sense antigenome is synthesized.

VI. Transcription at the gene-start site is not perfect, which leads to a gradient of mRNA abundance that decreases according to the distance from the 3' end of the genome.

Gradient of mRNA abundance from 3' end towards 5' end:

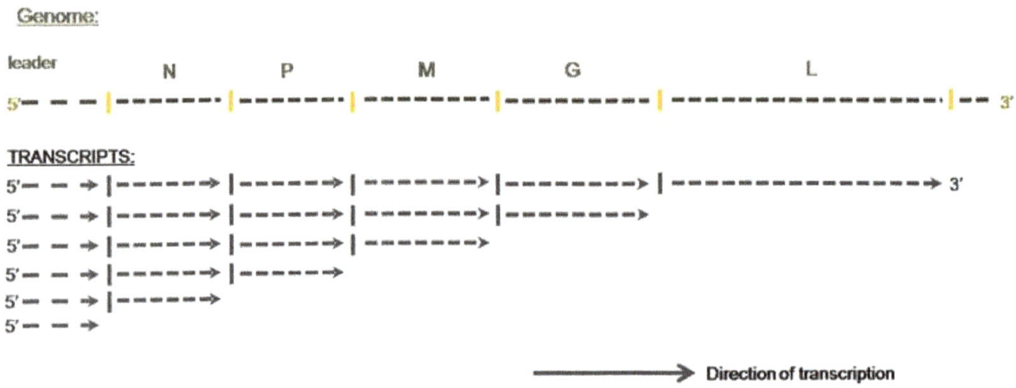

N= Nucleocapsid protein

P= Phosphoprotein

M= Matrix protein

G= Membrane glycoproteins

L= Large polymerase protein

Genome Replication

Virus enters the cell by receptor mediated endocytosis. The initial step after entering inside the cell is to transcribe viral mRNA from genomic RNA with the help of RDRP. The mRNA is further translated to form the viral proteins. The viral replication begins at 3' end and forms a complete positive sense RNA using negative sense genomic template. The viral transcription and replication occurs within a nucleocapsid-polymerase complex that consists of N, P, and L proteins. The switch from transcription to replication occurs when sufficient amount of N protein accumulates in the cytoplasm; it binds to the P protein to form a soluble complex, which is used for replication of the progeny RNA for the genome.

Replication strategy in negative sense RNA viruses:

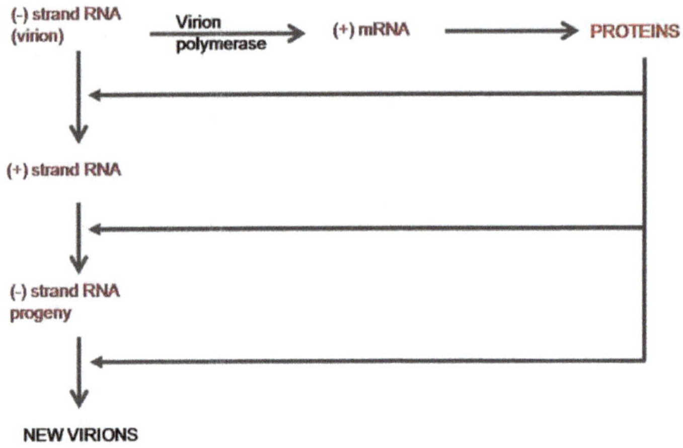

Paramyxoviruses

Paramyxoviruses are included under the family *Paramyxoviridae* , order *Mononegavirales* along with *Rhabdoviridae* , *Flioviridae* , and *Bornaviridae* . All the viruses in this group are enveloped and contain surface glycoproteins over it. The genome of the virus is single stranded RNA of negative polarity. Many life threatening diseases are caused by members of the family *Paramyxoviridae* . The impact of the disease caused by these viruses has been reduced dramatically through the use of vaccine.

Classification of Paramyxoviruses:

The family is divided into two subfamilies which are further divided into different genera.

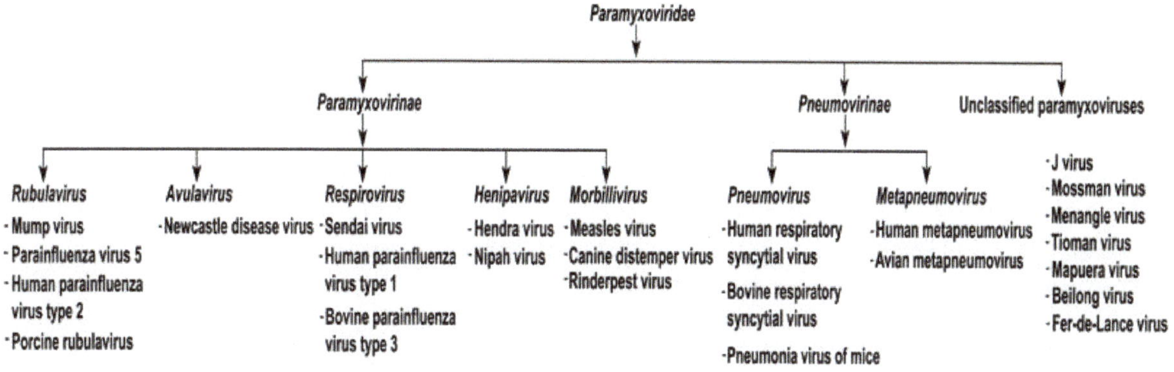

Virion Properties

Paramyxoviruses are pleomorphic and about 150- 350 nm in diameter. Virions are enveloped and covered by surface glycoproteins. The genome consists of ssRNA of negative polarity and 13-19 kb in size. The RNA at 3' end is not polyadenylated and 5' end of the RNA is not capped. With the exception of family *Pneumovirinae* , the genome size of the viruses are the even multiples (also called as rule of six) of six. The viral N protein binds effectively with the six nucleotide of the genomic RNA for its effective replication.

General Concepts

I. The Paramyxoviruses are the leading cause of respiratory disease in children; general illnesses include croup and inflammation of respiratory tract.

II. Paramyxoviruses share similar features; they contain a bilayer envelope containing spikes, have a helical symmetry and contain a negative stranded ssRNA genome. An RNA-dependent RNA polymerase (RDRP) is carried by the virus particle in order to perform the replication of the RNA genome.

III. Replication of the virus takes place in the cytoplasm and are released by the budding process.

IV. Virus antigens are confined in the lipid envelopes (spikes) and within the nucleocapsid core.

V. The viruses have a wide variety of host range that includes humans and primates.

VI. Viruses produce syncytia upon infection to susceptible cells by fusion and later cell lysis.

Different Paramyxoviruses:

Parainfluenza:

Nearly 25-35% of acute respiratory infections in infants and children are caused by this group of viruses. Disease starts with mild flu like symptoms which may progress to life-threatening (Croup, bronchiolitis and pneumonia) condition in the untreated cases. Parainfluenza viruses are the most common cause of croup. The viruses were divided into 4 distinct serotypes (numbered 1-4). These serotypes usually produce local inflammation in the upper and lower respiratory tract causing denudation of the ciliated epithelium (nose and throat). The virus generally sheds over 5-12 days following infection. Serotype 1 and 2 are attributed for the severe forms of the disease in young children.

Mumps:

This is another common disease of children. Acute infection of Mumps virus produces inflammation of salivary glands leading to its enlargement. Only one serotype is available for this group of virus. The common target tissues include glandular and nervous tissue. The virus enters through the pharynx or conjunctiva, s ystemic infection of the virus can cause viremia. Secondary dissemination of the virus occurs to salivary glands, gonads, pancreas, and central nervous system after their multiplication in the lymphoid tissues. Incubation period of the disease may vary between 18-24 days while in many cases it is asymptomatic. The most characteristic feature of the disease is painful swelling of the parotid glands. Sometimes disease may lead to deafness and severe inflammation in the male reproductive system.

Measles:

It is also an acute disease of infants and children. The virus commonly causes a rash over the body with a high fever, occasionally conjunctivitis and pneumonia. In severe form of the disease virus may cause inflammation and pathological condition in the brain. Like Mumps virus only one serotype exists for Measles virus. Measles is also a systemic infection spread by dissemination of the virus through blood. Acute disease affects the lymphatic and respiratory systems while persistence of the virus in children leads to subacute sclerosing panencephalitis . Virus enters the body via the oropharyngeal route, multiplies locally within the lymphatic system, and spreads to the mucosal surface of respiratory, gastrointestinal and central nervous system. Clinically, respiratory symptoms and fever are evident during the early stages which on later changes to a rash during the eruptive phase. Rash over the body and head are sometimes called as Koplik's spots . They are ulcerated mucosal lesions characterized by necrosis and infiltration of neutrophils, and are the pathognomonic (hallmark and unique to measles) signs of the measles. Secondary infection by bacteria may sometimes complicate the situation and even turn worse in untreated condition.

Respiratory Syncytial Virus :

Respiratory Syncytial virus is one of the leading causes of bronchiolitis and pneumonia in infants under one year of age. The viruses produce a characteristic syncytia formation in the respiratory epithelium cells; hence the name is given as respiratory syncytial virus. The virus starts its infection in the upper or lower respiratory tract infecting ciliated epithelium. Spread of the virus proceeds by cell fusion. Severe form of the disease may cause bronchiolitis, pneumonia, or croup in infants.

Croup is sometimes called as barking cough and is characterized by swelling around the vocal cords. It usually associated with the inflammation of larynx, trachea, and bronchioles.

Orthomyxoviruses

The name originates from the Greek word “ortho” which means correct while “myxo” stand for mucus . The word essentially stands for virus that infects epithelial cells in the right way, exactly opposite to that of paramyxoviruses. The family *Orthomyxoviridae* contains viruses of single stranded segmented RNA genome (6-8 segments). Out of all, influenza viruses are the most important members of this family which includes influenza virus A, B, and C. The name originates during 18 th century from a disease that was thought to “ influence” by stars. The pathogenic viruses are included in the genus influenza virus A, whereas other two genera (B and C) circulate constantly in human subjects. Several devastating pandemics caused because of influenza virus in past include famous Spanish Flu (1918), Asian Flu (1956), and Hong Kong Flu (1967) which killed millions of people.

Distinct Characters

I. The viruses cause an acute respiratory disease with prominent systemic symptoms with its major manifestation on the respiratory system.

II. Influenza virus type A is responsible for periodic epidemics worldwide; while virus types A and B cause regional epidemics during the cold weather.

III. Antigenic drift (minor changes in the viral surface proteins) and Antigenic Shift (major changes in the viral genome due to rearrangement of the virus segments or reassortments) are responsible for both epidemics and pandemics of influenza viruses.

Table Differences between influenza virus types:

	Influenza virus type A	Influenza virus type B	Influenza virus type C
Animal Reservoir	Present	Absent	Absent
Spread	Pandemic/ Epidemic	Epidemic	Sporadic
Severity	High	Moderate	Mild
Variations	Antigenic shift/drift	Antigenic drift	Antigenic drift

Virus Classification

The family *Orthomyxoviridae* contains the genera Influenza virus A, Influenza virus B, Influenza virus C, Thogotovirus, and Isavirus.

Different members of family *Orthomyxoviridae* :

Genus	Virus type
Influenzavirus A	Influenza A virus
Influenzavirus B	Influenza B virus
Influenzavirus C	Influenza C virus
Thogotovirus	Thogotovirus
	Dhori virus
	Batken virus
Isavirus	Infectious salmon anemia virus

Virion Properties

Orthomyxovirus virions are pleomorphic in shape and around 80–120 nm in diameter. The nucleocapsids are helical in symmetry and contains eight (influenza virus A, B, and isavirus), seven (influenza virus C), or six (Thogotovirus) RNA segments. Genome is single stranded negative sense RNA of 10 – 14Kb in size. The viruses also contain surface glycoproteins over the lipid envelope. The two envelope glycoproteins are hemagglutinin protein (H) and neuraminidase protein (N). There are 15 different subtypes of "H" and 9 different subtypes of "N". This provides a total of 135 (15 x 9) possible combinations. Hemagglutinin is the most important protein of the virus and determinant of its virulence while neuraminidase helps in the budding and release of the progeny virions. Both hemagglutinin and neuraminidase frequently undergo genetic modifications decreasing the effectiveness of the host immune response. Influenza virus "C" lacks the "N" protein. Envelope is lined by Matrix protein "M1" and an ion channel matrix protein "M2". Three proteins namely PB1, PB2 and PA form the viral RNA polymerase complex which is associated with genomic RNA and nucleoprotein.

Schematic representation of an influenza virus

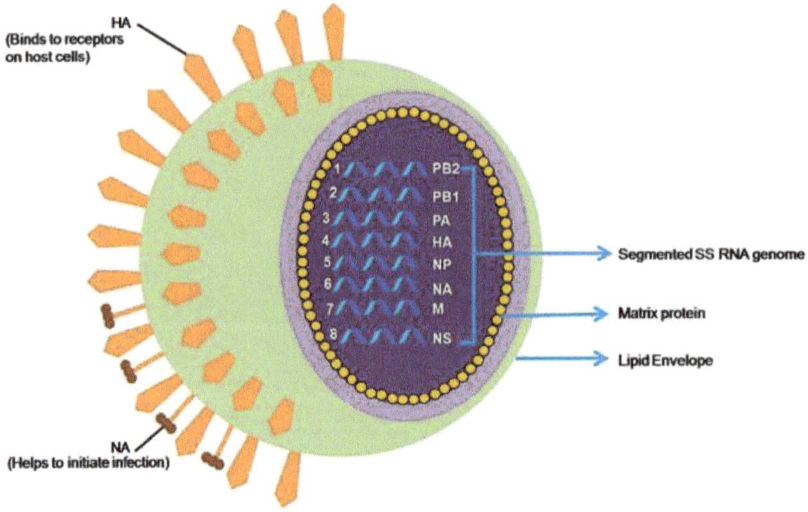

Virus Replication

Influenza virus enters the cell after binding to sialic acid receptor present on the cell membrane. Different cells of the body contain different sialic acid receptor types which largely determine the host range of these viruses. The respiratory tract epithelium of humans contains α 2-6 linkage sialic acid receptor while birds contain α 2-3 sialic acid receptor. The types of sialic acid receptor in the epithelium determines the tropism of human as well as avian influenza viruses. A single amino acid change in the hemagglutinin protein [E (Glutamic acid) 190D (Aspartic acid)] changed the binding efficiency of influenza virus from α2,3 to α2,6 which caused the outbreak of 1918 influenza virus (Spanish flu). Virus enters the cell by receptor mediated endocytosis and uncoats under the low pH condition of endosome. RNA synthesis of influenza virus takes place in the nucleus of the cell. Nucleoprotein of influenza virus contains nuclear localization signals (NLS) that help in transportation of ribonucleoprotein complex into the nucleus. Negative sense RNA genome of influenza viruses serve as a template for the synthesis of positive sense RNA. Positive sense replicative intermediate RNA acts as a template for progeny RNA genome. Interestingly, viral endonuclease activity of PB2 protein cleaves the 5' cap and 10-13 nucleotides from a cellular mRNA in order to transcribe the viral RNA. The phenomenon is called as cap snatching . All orthomyxoviruses undergo splicing phenomena to produce two proteins from one gene such as influenza virus A uses gene segment 7 to produce M1 and M2 protein. Similarly 8 th segment of influenza virus produces NS1 and NS2 protein after undergoing splicing. In certain influenza viruses, frame shift mutation leads to formation of PB1-F2 protein. Viral protein synthesis occurs in the cytoplasm and its maturation takes place in endoplasmic reticulum and Golgi apparatus. Viral nucleoproteins are required for replication of genomic RNA which then enters the nucleus along with polymerase protein for transcription. Progeny virus is released by budding through the plasma membrane.

Schematic representation of an influenza virus replication cycle

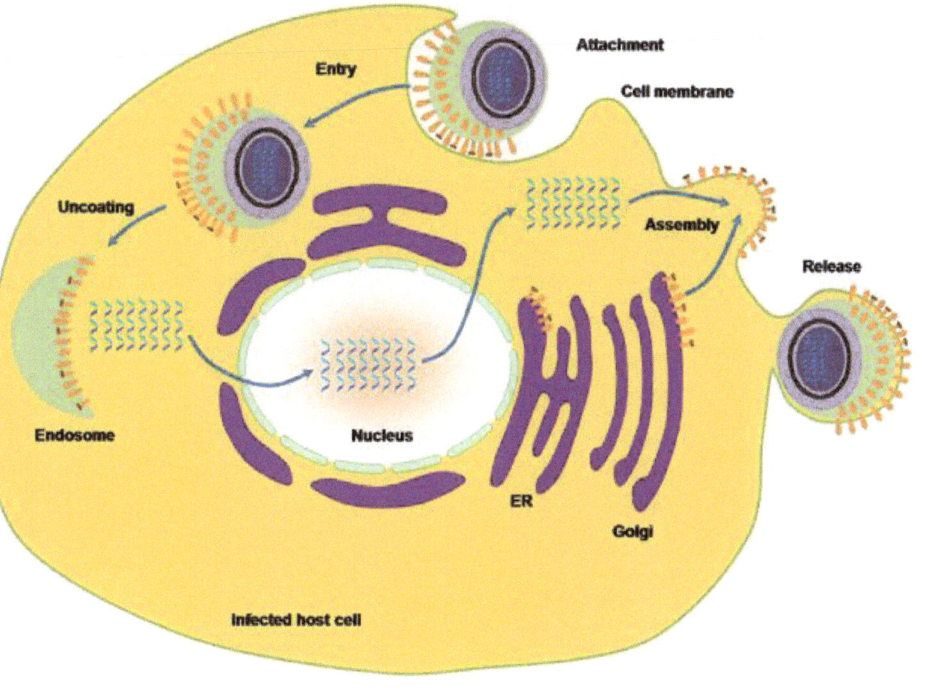

Influenza Pathogenesis

When influenza virus (commonly referred as flu) enters the host via aerosol it replicates in the epithelial cells of upper and lower respiratory tract. The infection causes destruction of the ciliated cells along the lining of the respiratory tract that leads to inflammation and formation of exudates. The signs include tracheitis, bronchitis, and pneumonitis. Sometime collapse of adjacent air sacs and emphysema is also evident. Microscopically, there is complete denudation of the epithelial lining and infiltration of inflammatory cells.

Mode of entry and exit of influenza virus in humans

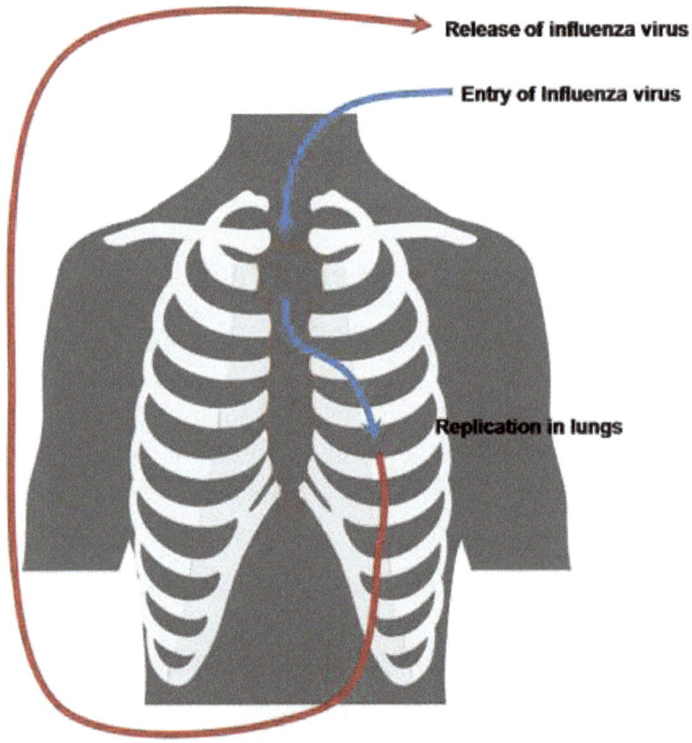

Molecular Determinants

The hemagglutinin (HA) protein of influenza virus is a key in determining the virulence and pathogenicity of influenza virus. HA protein of influenza virus has to cleave in order to start the infectious cycle. The highly pathogenic influenza virus has many basic amino acid (arginine) residues at the cleavage site of the HA protein, while the less pathogenic viruses contains less basic amino acid residues at the cleavage site. The virus containing less basic amino acids are restricted to multiply in certain selected tissues as the protease required for the cleavage site is present only in certain specific tissues. In humans and birds respiratory and gastrointestinal tract contains the proteases to cleave the HA protein. In contrast, the viruses containing more basic amino acid residues at the cleavage site increases the range of cells capable of producing infectious virions. This property of the virus is important in determining its tissue tropism.

Non structural (NS1) protein in influenza virus is another key factor in determining the virulence of the viral strains. NS1 protein binds to dsRNA and blocks the interferon activation. The influenza virus containing mutation in the NS1 proteins are attenuated as the mutated form of the protein does not bind with the dsRNA which causes increased level of interferon in the infected cells.

Polymerase protein PB2 is another important component that determines the virulence of influenza viruses. PB2 helps in RNA transcription and replication of the virus. The PB2 containing mutation in certain amino acid sequences may increase or decrease the virulence of the virus. Another protein such as PB1 is also suggested as a determinant of influenza virus virulence.

Clinical Features

The incubation period of the virus in human is from 2 to 4 days, may extend up to 8 days based on the immune status of the patient. Initial symptoms of the disease include fever and lower respiratory tract illness. Diarrhea, abdominal pain, headache, and vomiting are other common symptoms observed during the early phase of infection. The later phase of the disease shows the symptoms of more advanced respiratory distress such as acute respiratory distress syndrome (ARDS). Multi-organ failure involving kidney and heart are also associated during the later part of the disease.

Influenza Virus Reassortment

Reassortment is another very interesting phenomenon and occurs in viruses having segmented genome. Reassortment stands for mixing of viral genetic information in order to evolve them as a new species having potential to cause a greater pandemic. If a cell or animal is infected with two different influenza viruses, the RNA segments of both viruses are copied in the nucleus. When new virus particles are assembled at the cell membrane, each of the 8 RNA segments may originate from any one of them or a mixture of two infecting viruses. The progeny virion that contains RNA segments from both the parents is called as reassortant. This process is illustrated in the diagram below for involution of swine influenza virus.

Reassortment of avian and human influenza virus in swine

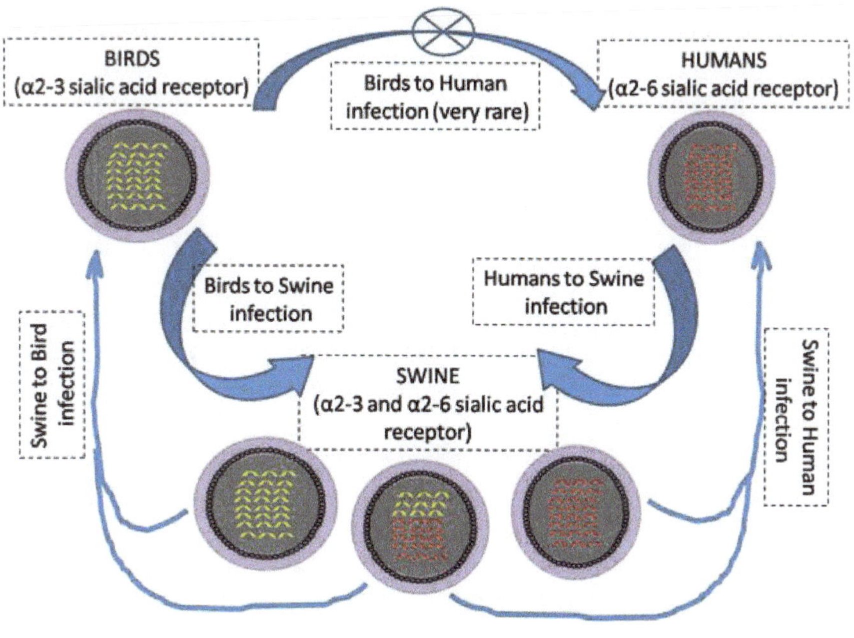

Avian Influenza

Avian influenza — known informally as *avian flu* or bird flu — refers to "influenza caused by viruses adapted to birds." The type with the greatest risk is highly pathogenic avian influenza (HPAI).

"Bird flu" is a phrase similar to "swine flu," "dog flu," "horse flu," or "human flu" in that it refers to an illness caused by any of many different strains of influenza viruses that have adapted to a specific host. Out of the three types of influenza viruses (A, B, and C), influenza A virus is a zoonotic infection with a natural reservoir almost entirely in birds. Avian influenza, for most purposes, refers to the influenza A virus.

Though influenza A is adapted to birds, it can also stably adapt and sustain person-to person transmission. Recent influenza research into the genes of the Spanish flu virus shows it to have genes adapted from both human and avian strains. Pigs can also be infected with human, avian, and swine influenza viruses, allow for mixtures of genes (reassortment) to create a new virus, which can cause an antigenic shift to a new influenza A virus subtype which most people have little to no immune protection.

Avian influenza strains are divided into two types based on their pathogenicity: high pathogenicity (HP) or low pathogenicity (LP). The most well-known HPAI strain, H5N1, appeared in China in 1996, and also has low pathogenic strains found in North America. Companion birds in captivity are unlikely to contract the virus and there has been no report of a companion bird with avian influenza since 2003. Pigeons do not contract or spread the virus.

In the approximate four years from early 2013 to early 2017, 916 lab-confirmed human cases of H7N9 were reported to the World Health Organization (WHO).

On 9 January 2017, the National Health and Family Planning Commission of China reported to WHO 106 cases of H7N9 which occurred from late November through late December, including 35 deaths, 2 potential cases of human-to-human transmission, and 80 of these 106 persons stating that they have visited live poultry markets. The cases are reported from Jiangsu (52), Zhejiang (21), Anhui (14), Guangdong (14), Shanghai (2), Fujian (2) and Hunan (1). Similar sudden increases in the number of human cases of H7N9 have occurred in previous years during December and January.

History

The most widely quoted date for the beginning of recorded history of avian influenza (initially known as fowl plague) was in 1878 when it was differentiated from other diseases that caused high mortality rates in birds. Fowl plague, however, also included Newcastle disease until as recently as the 1950s. Between 1959 and 1995, there were 15 recorded occasions of the emergence of HPAI viruses in poultry, but losses were minimal. Between 1996 and 2008 however, HPAI outbreaks in poultry have occurred at least 11 times and 4 of these outbreaks have involved millions of birds.

In the 1990s, the world's poultry population grew 76% in developing countries and 23% in developed countries, contributing to the increased prevalence of avian influenza. Before the 1990s, HPAI caused high mortality in poultry, but infections were sporadic and contained. Outbreaks have become more common due to the high density and frequent movement of flocks from intensive poultry production.

Influenza A/H5N1 was first isolated from a goose in China in 1996. Human infections were first reported in 1997 in Hong Kong. Since 2003, more than 700 human cases of Asian HPAI H5N1 have been reported to the WHO, primarily from 15 countries in Asia, Africa, the Pacific, Europe, and the Middle East, though over 60 countries have been affected.

Genetics

Genetic factors in distinguishing between "human flu viruses" and "avian flu viruses" include:

- PB2: (RNA polymerase): Amino acid (or residue) position 627 in the PB2 protein encoded by the PB2 RNA gene. Until H5N1, all known avian influenza viruses had a Glu at position 627, while all human influenza viruses had a Lys.
- HA: (hemagglutinin): Avian influenza HA viruses bind alpha 2-3 sialic acid receptors, while human influenza HA viruses bind alpha 2-6 sialic acid receptors. Swine influenza viruses have the ability to bind both types of sialic acid receptors. Hemagglutinin is the major antigen of the virus against which neutralizing antibodies are produced, and influenza virus epidemics are associated with changes in its antigenic structure. This was originally derived from pigs, and should technically be referred to as "pig flu".

Subtypes

There are many subtypes of avian influenza viruses, but only some strains of five subtypes have been known to infect humans: H5N1, H7N3, H7N7, H7N9, and H9N2. At least one person, an elderly woman in Jiangxi Province, China, died of pneumonia in December 2013 from the H10N8 strain, the first human fatality confirmed to be caused by that strain.

Most human cases of the avian flu are a result of either handling dead infected birds or from contact with infected fluids. It can also be spread through contaminated surfaces and droppings. While most wild birds mainly have only a mild form of the H5N1 strain, once domesticated birds such as chickens or turkeys are infected, it can become much more deadly because the birds are often in close contact. There is currently a large threat of this in Asia with infected poultry due to low hygiene conditions and close quarters. Although it is easy for humans to contract the infection from birds, human-to-human transmission is more difficult without prolonged contact. However, public health officials are concerned that strains of avian flu may mutate to become easily transmissible between humans.

Spreading of H5N1 from Asia to Europe is much more likely caused by both legal and illegal poultry trades than dispersing through wild bird migrations, being that in recent studies, there were no secondary rises in infection in Asia when wild birds migrate south again from their breeding grounds. Instead, the infection patterns followed transportation such as railroads, roads, and country borders, suggesting poultry trade as being much more likely. While there have been strains of avian flu to exist in the United States, such as Texas in 2004, they have been extinguished and have not been known to infect humans.

Examples of avian influenza A virus strains:

HA subtype designation	NA subtype designation	Avian influenza A viruses
H1	N1	A/duck/Alberta/35/76(H1N1)
H1	N8	A/duck/Alberta/97/77(H1N8)
H2	N9	A/duck/Germany/1/72(H2N9)
H3	N8	A/duck/Ukraine/63(H3N8)
H3	N8	A/duck/England/62(H3N8)
H3	N2	A/turkey/England/69(H3N2)
H4	N6	A/duck/Czechoslovakia/56(H4N6)
H4	N3	A/duck/Alberta/300/77(H4N3)
H5	N3	A/tern/South Africa/300/77(H4N3)
H5	N4	A/Ethiopia/300/77(H6N6)
H5	N8	H5N8
H5	N9	A/turkey/Ontario/7732/66(H5N9)
H5	N1	A/chick/Scotland/59(H5N1)
H6	N2	A/turkey/Massachusetts/3740/65(H6N2)
H6	N8	A/turkey/Canada/63(H6N8)
H6	N5	A/shearwater/Australia/72(H6N5)
H6	N6	A/jyotichinara/Ehiopia/73(H6N6)
H6	N1	A/duck/Germany/1868/68(H6N1)
H7	N7	A/fowl plague virus/Dutch/27(H7N7)
H7	N1	A/chick/Brescia/1902(H7N1)
H7	N9	A/chick/China/2013(H7N9)
H7	N3	A/turkey/England/639H7N3)
H7	N1	A/fowl plague virus/Rostock/34(H7N1)
H8	N4	A/turkey/Ontario/6118/68(H8N4)
H9	N2	A/turkey/Wisconsin/1/66(H9N2)
H9	N6	A/duck/Hong Kong/147/77(H9N6)
H9	N6	A/duck/Hong Kong/147/77(H9N6)
H9	N8	A/manishsurpur/Malawi/149/77(H9N8)
H9	N7	A/turkey/Scotland/70(H9N7)
H10	N8	A/quail/Italy/1117/65(H10N8)
H11	N6	A/duck/England/56(H11N6)
H11	N9	A/duck/Memphis/546/74(H11N9)
H12	N5	A/duck/Alberta/60/76/(H12N5)
H13	N6	A/gull/Maryland/704/77(H13N6)
H14	N4	A/duck/Gurjev/263/83(H14N4)
H15	N9	A/shearwater/Australia/2576/83(H15N9)

Spread

Avian influenza is most often spread by contact between infected and healthy birds, though can also be spread indirectly through contaminated equipment. The virus is found in secretions from the nostrils, mouth, and eyes of infected birds as well as their droppings. HPAI infection is spread to people often through direct contact with infected poultry, such as during slaughter or plucking. Though the virus can spread through airborne secretions, the disease itself is not an airborne disease. Highly pathogenic strains spread quickly among flocks and can destroy a flock within 28 hours; the less pathogenic strains may affect egg production but are much less deadly.

Although it is possible for humans to contract the avian influenza virus from birds, human-to-human contact is much more difficult without prolonged contact. However, public health officials are concerned that strains of avian flu may mutate to become easily transmissible between humans. Some strains of avian influenza are present in the intestinal tract of large numbers of shore birds and water birds, but these strains rarely cause human infection.

Five manmade ecosystems have contributed to modern avian influenza virus ecology: integrated indoor commercial poultry, range-raised commercial poultry, live poultry markets, backyard and hobby flocks, and bird collection and trading systems including cockfighting. Indoor commercial poultry has had the largest impact on the spread of HPAI, with the increase in HPAI outbreaks largely the result of increased commercial production since the 1990s.

Village Poultry

In the early days of the HPAI H5N1 pandemic, village poultry and their owners were frequently implicated in disease transmission. Village poultry, also known as backyard and hobby flocks, are small flocks raised under extensive conditions and often allowed free range between multiple households. However, research has shown that these flocks pose less of a threat than intensively raised commercial poultry with homogenous genetic stock and poor biosecurity. Backyard and village poultry also do not travel great distances compared to transport of intensively raised poultry and contribute less to the spread of HPAI. This initial implication of Asian poultry farmers as one broad category presented challenges to prevention recommendations as commercial strategies did not necessarily apply to backyard poultry flocks.

H5N1

H5N1

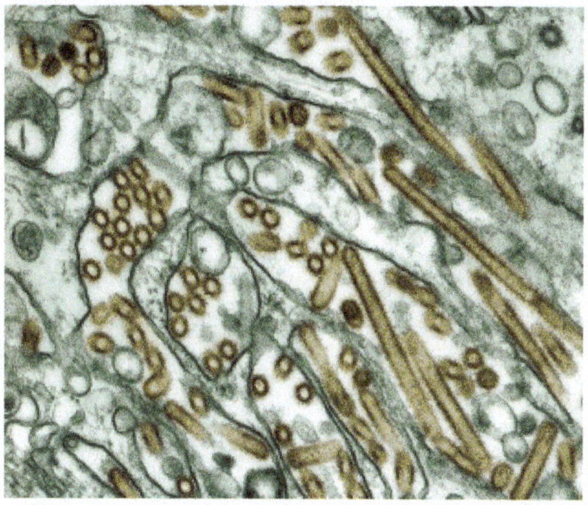

- Influenza A virus
    - subtype H_5N_1
- Genetic structure
- Infection
- Human mortality
- Global spread
    - in 2004
    - in 2005
    - in 2006
    - in 2007
- Social impact
- Pandemic
- Vaccine

The highly pathogenic influenza A virus subtype H_5N_1 is an emerging avian influenza virus that is causing global concern as a potential pandemic threat. It is often referred to simply as "bird flu" or "avian influenza", even though it is only one of many subtypes.

H_5N_1 has killed millions of poultry in a growing number of countries throughout Asia, Europe, and Africa. Health experts are concerned that the coexistence of human flu viruses and avian flu viruses (especially H_5N_1) will provide an opportunity for genetic material to be exchanged between species-specific viruses, possibly creating a new virulent influenza strain that is easily transmissible and lethal to humans. The mortality rate for humans with H_5N_1 is 60%.

Since first human H_5N_1 outbreak occurred in 1997, there has been an increasing number of HPAI H_5N_1 bird-to-human transmissions, leading to clinically severe and fatal human infections. Because a significant species barrier exists between birds and humans, the virus does not easily spread to humans, however some cases of infection are being researched to discern whether human-to-human transmission is occurring. More research is necessary to understand the pathogenesis and epidemiology of the H_5N_1 virus in humans. Exposure routes and other disease transmission characteristics, such as genetic and immunological factors that may increase the likelihood of infection, are not clearly understood.

The first known transmission of H_5N_1 to a human occurred in Hong Kong in 1997, when there was an outbreak of 18 human cases; 6 deaths were confirmed. None of the infected people worked with poultry. After culling all of the poultry in the area, no more cases were diagnosed. In 2006, the first human-to-human transmission likely occurred when 7 members of a family in Sumatra became infected after contact with a family member who had worked with infected poultry.

Although millions of birds have become infected with the virus since its discovery, 359 people have died from the H_5N_1 in twelve countries according to World Health Organization reports as of August 10, 2012.

As an example, the H_5N_1 outbreak in Thailand caused massive economic losses, especially among poultry workers. Infected birds were culled and slaughtered. The public lost confidence with the poultry products, thus decreasing the consumption of chicken products. This also elicited a ban from importing countries. There were, however, factors which aggravated the spread of the virus, including bird migration, cool temperature (increases virus survival) and several festivals at that time.

A mutation in the virus was discovered in two Guangdong patients in February 2017 which rendered it more deadly to chickens, inasmuch as it could infect every organ; the risk to humans was not increased, however.

Controversial Research

A study published in 2012 in Science Magazine reported on research findings that allowed for the airborne transmission of H_5N_1 in laboratory ferrets. The study identified the 5 mutations necessary for the virus to become airborne and immediately spiked controversy over the ethical implications of making such potentially dangerous information available to the general public. The study was allowed to remain available in its entirety, though it remains a controversial topic within the scientific community.

The study in question, however, created airborne H_5N_1 via amino acid substitutions that largely mitigated the devastating effects of the disease. This fact was underscored by the 0% fatality rate among the ferrets infected via airborne transmission, as well as the fundamental biology underlying the substitutions. Flu viruses attach to host cells via the hemagluttinin proteins on their envelope. These hemagluttinin proteins bind to sialic acid receptors on host cells, which can fall into two categories. The sialic acid receptors can be either 2,3 or 2,6-linked, with the species of origin largely deciding receptor preference. In influenzas of avian origin 2,3-linkage is preferred, vs. influenzas of human origin in which 2,6-linkage is preferable. 2,3-linked SA receptors in humans are found predominantly in the lower respiratory tract, a fact that is the primary foundation for the deadliness of avian influenzas in humans, and also the key to their lack of airborne transmission. In the study that created an airborne avian influenza among ferrets it was necessary to switch the receptor preference of the host cells to those of 2,6-linkage, found predominantly in humans' upper respiratory tract, in order to create an infection that could shed aerosolized virus particles. Such an infection, however, must occur in the upper respiratory tract of humans, thus fundamentally undercutting the fatal trajectory of the disease.

H_7N_9

Influenza A virus subtype H_7N_9 is a novel avian influenza virus first reported to have infected humans in 2013 in China. Most of the reported cases of human infection have resulted in severe respiratory illness. In the month following the report of the first case, more than 100 people had been infected, an unusually high rate for a new infection; a fifth of those patients had died, a fifth had recovered, and the rest remained critically ill. The World Health Organization (WHO) has identified H_7N_9 as "...an unusually dangerous virus for humans." As of June 30, 133 cases have been reported, resulting in the deaths of 43.

Research regarding background and transmission is ongoing. It has been established that many of the human cases of H_7N_9 appear to have a link to live bird markets. As of July 2013, there had been no evidence of sustained human-to-human transmission, however a study group headed by one of the world's leading experts on avian flu reported that several instances of human-to-human infection were suspected. It has been reported that H7N9 virus does not kill poultry, which will make surveillance much more difficult. Researchers have commented on the unusual prevalence of older males among H7N9-infected patients. While several environmental, behavioral, and biological explanations for this pattern have been proposed, as yet, the reason is unknown. Currently no vaccine exists, but the use of influenza antiviral drugs known as neuraminidase inhibitors in cases of early infection may be effective.

The number of cases detected after April fell abruptly. The decrease in the number of new human H7N9 cases may have resulted from containment measures taken by Chinese authorities, including closing live bird markets, or from a change in seasons, or possibly a combination of both factors. Studies indicate that avian influenza viruses have a seasonal pattern, thus it is thought that infections may pick up again when the weather turns cooler in China.

In the four years from early 2013 to early 2017, 916 lab-confirmed human cases of H7N9 were reported to WHO.

On 9 January 2017, the National Health and Family Planning Commission of China reported to WHO 106 cases which occurred from late November through December. 29, 2016. The cases are reported from Jiangsu (52), Zhejiang (21), Anhui (14), Guangdong (14), Shanghai (2), Fujian (2) and Hunan (1). 80 of these 106 persons have visited live poultry markets. Of these cases, there have been 35 deaths. In two of the 106 cases, human-to-human transmission could not be ruled out.

Affected prefectures in Jiangsu province closed live poultry markets in late December 2016, whereas Zhejiang, Guangdong and Anhui provinces went the route of strengthening live poultry market regulations. Travellers to affected regions are recommended to avoid poultry farms, live bird markets, and surfaces which appear to be contaminated with poultry feces [shit]. Similar sudden increases in the number of human cases of H7H9 have occurred in previous years during December and January.

In Domestic Animals

Several domestic species have been infected with and shown symptoms of H5N1 viral infection, including cats, dogs, ferrets, pigs, and birds.

Birds

Attempts are made in the United States to minimize the presence of HPAI in poultry through routine surveillance of poultry flocks in commercial poultry operations. Detection of a HPAI virus may result in immediate culling of the flock. Less pathogenic viruses are controlled by vaccination, which is done primarily in turkey flocks (ATCvet codes: QI01AA23 (WHO) for the inactivated fowl vaccine, QI01CL01 (WHO) for the inactivated turkey combination vaccine).

Seals

A recent strain of the virus is able to infect the lungs of seals.

Cats

Avian influenza in cats can show a variety of symptoms and usually lead to death. Cats are able to get infected by either consuming an infected bird or by contracting the virus from another infected cat.

Global Impact

In 2005, the formation of the International Partnership on Avian and Pandemic Influenza was announced in order to elevate the importance of avian flu, coordinate efforts, and improve disease reporting and surveillance in order to better respond to future pandemics. New networks of laboratories have emerged to detect and respond to avian flu, such as the Crisis Management Center for Animal Health, the Global Avian Influenza Network for Surveillance, OFFLU, and the Global Early Warning System for major animal diseases. After the 2003 outbreak, WHO member states have also recognized the need for more transparent and equitable sharing of vaccines and other benefits from these networks. Cooperative measures created in response to HPAI have served as a basis for programs related to other emerging and re-emerging infectious diseases.

HPAI control has also been used for political ends. In Indonesia, negotiations with global response networks were used to recentralize power and funding to the Ministry of Health. In Vietnam policymakers, with the support of the Food and Agriculture Organization of the United Nations (FAO), used HPAI control to accelerate the industrialization of livestock production for export by proposing to increase the portion of large-scale commercial farms and reducing the number of poultry keepers from 8 to 2 million by 2010.

Stigma

Backyard poultry production was viewed as "traditional Asian" agricultural practices that contrasted with modern commercial poultry production and seen as a threat to biosecurity. Backyard production appeared to hold greater risk than commercial production due to lack of biosecurity and close contact with humans, though HPAI spread in intensively raised flocks was greater due to high density rearing and genetic homogeneity. Asian culture itself was blamed as the reason why certain interventions, such as those that only looked at placed-based interventions, would fail without looking for a multifaceted solutions.

Indonesia

Press accounts of avian flu in Indonesia were seen by poultry farmers as conflating suspected cases while the public did see the accounts as informative, though many became de-sensitized to the idea of impending danger or only temporarily changed their poultry-related behavior. Rumors also circulated in Java in 2006. These tended to focus on bird flu being linked to big businesses in order to drive small farmers out of the market by exaggerating the danger of avian influenza, avian flu being introduced by foreigners to force Indonesians to purchase imported chicken and keep Indonesian chicken off the world market, and the government using avian flu as a ploy to attract funds from wealthy countries. Such rumors reflected concerns about big businesses, globalization, and a distrust of the national government in a country where "the amount of decentralization here is breathtaking" according to Steven Bjorge, a WHO epidemiologist in Jakarta in 2006.

In the context a decentralized national government that the public did not completely trust, Indonesian Health Minister Siti Fadilah Supari announced in December 2006 that her government would no longer be sharing samples of H5N1 collected from Indonesian patients. This decision came as a shock to the international community as it disrupted the Global Influenza Surveillance Network (GISN) coordinated by the WHO for managing seasonal and pandemic influenza. GISN is based on countries sharing virus specimens freely with the WHO which assesses and eventually sends these samples to pharmaceutical companies in order to produce vaccines that are sold back to these countries. Though this was initially seen as an attempt to protect national sovereignty at all costs, it was instead used for a domestic political struggle. Prior to Indonesia's dispute with the GISN, the Ministry of Health, already weak due to the decentralized nature the government, was experiencing further leakage of funding to state and non-state agencies due to global health interventions. By reasserting control over public health issues and funding by setting itself up as the sole Indonesian representative to the WHO, the Ministry of Health made itself a key player in the management of future international funds relating vaccine production and renegotiated benefits from global surveillance networks.

Economic

Approximately 20% of the protein consumed in developing countries come from poultry. In the wake of the H5N1 pandemic, millions of poultry were killed. In Vietnam alone, over 50 million domestic birds were killed due to HPAI infection and control attempts. A 2005 report by the FAO totaled economic losses in South East Asia around US$10 billion. This had the greatest impact on small scale commercial and backyard producers relative to total assets compared to industrial chains which primarily experience temporary decreases in exports and loss of consumer confidence. Some governments did provide compensation for culled poultry, it was often far below market value (close to 30% of market value in Vietnam), while others such as Cambodia provide no compensation to farmers at all.

As poultry serves as a source of food security and liquid assets, the most vulnerable populations were poor small scale farmers. The loss of birds due to HPAI and culling in Vietnam led to an average loss of 2.3 months of production and US$69–108 for households where many have an income of $2 a day or less. The loss of food security for vulnerable households can be seen in the stunting of children under 5 in Egypt. Women are another population at risk as in most regions of the world, small flocks are tended to by women. Widespread culling also resulted in the decreased enrollment of girls in school in Turkey.

Prevention

People who do not regularly come into contact with birds are not at high risk for contracting avian influenza. Those at high risk include poultry farm workers, animal control workers, wildlife biologists, and ornithologists who handle live birds. Organizations with high-risk workers should have an avian influenza response plan in place before any cases have been discovered. Biosecurity of poultry flocks is also important for prevention. Flocks should be isolated from outside birds, especially wild birds, and their waste; vehicles used around the flock should be regularly disinfected and not shared between farms; and birds from slaughter channels should not be returned to the farm.

With proper infection control and use of personal protective equipment (PPE), the chance for infection is low. Protecting the eyes, nose, mouth, and hands is important for prevention because these are the most common ways for the virus to enter the body. Appropriate personal protective equipment includes aprons or coveralls, gloves, boots or boot covers, and a head cover or hair cover. Disposable PPE is recommended. An N-95 respirator and unvented/indirectly vented safety goggles are also part of appropriate PPE. A powered air purifying respirator (PAPR) with hood or helmet and face shield is also an option.

Proper reporting of an isolated case can help to prevent spread. The Centers for Disease Control and Prevention (US) recommendation is that if a worker develops symptoms within 10 days of working with infected poultry or potentially contaminated materials, they should seek care and notify their employer, who should notify public health officials.

For future avian influenza threats, the WHO suggests a 3 phase, 5 part plan.

- Phase: Pre-pandemic
    - Reduce opportunities for human infection
    - Strengthen the early warning system
- Phase: Emergence of a pandemic virus
    - Contain or delay spread at the source
- Phase: Pandemic declared and spreading internationally
    - Reduce morbidity, mortality, and social disruption
    - Conduct research to guide response measures

Vaccines for poultry have been formulated against several of the avian H5N1 influenza varieties. Control measures for HPAI encourage mass vaccinations of poultry though The World Health Organization has compiled a list of known clinical trials of pandemic influenza prototype vaccines, including those against H5N1. In some countries still at high risk for HPAI spread, there is compulsory strategic vaccination though vaccine supply shortages remain a problem.

For Village Poultry Farmers

Initial response to H5N1, a one size fits all recommendation was used for all poultry production systems, though measures for intensively raised birds were not necessarily appropriate for extensively raised birds. When looking at village poultry, it was first assumed that the household was the unit and that flocks did not make contact with other flocks, though more effective measures came into use when the epidemiological unit was the village.

Recommendations also involve restructuring commercial markets to improve biosecurity against avian influenza. Poultry production zoning is used to limit poultry farming to specific areas outside of urban environments while live poultry markets improve biosecurity by limiting the number of traders holding licenses and subjecting producers and traders to more stringent inspections. These recommendations in combination with requirements to fence and house all poultry, and limit free ranging flocks will eventually lead to fewer small commercial producers and backyard producers, costing livelihoods as they are unable to meet the conditions needed to participate.

A summary of reports to the World Organisation for Animal Health in 2005 and 2010 suggest that surveillance and under-reporting in developed and developing countries is still a challenge. Often, donor support can focus on HPAI control alone, while similar diseases such as Newcastle disease, acute fowl cholera, infectious laryngotracheitis, and infectious bursal disease still affect poultry populations. When HPAI tests come back negative, a lack of funded testing for differential diagnoses can leave farmers wondering what killed their birds.

Since traditional production systems require little investment and serve as a safety net for lower income households, prevention and treatment can been seen as less cost effective than letting a few birds die. Effective control not only requires prior agreements to be made with relevant government agencies, such as seen with Indonesia, they must also not unduly threaten food security.

Culling

Culling is used in order to decrease the threat of avian influenza transmission by killing potentially infected birds. The FAO manual on HPAI control recommends a zoning strategy which begins with the identification of an infected area (IA) where sick or dead birds have tested positive. All poultry in this zone are culled while the area 1 to 5 km from the outer boundary of the IA is considered the restricted area (RA) placed under strict surveillance. 2 to 10 km from the RA is the control area (CA) that serves as a buffer zone in case of spread. Culling is not recommended beyond the IA unless there is evidence of spread. The manual, however, also provides examples of how control was carried out between 2004 and 2005 to contain H5N1 where all poultry was to be stamped out in a 3 km radius beyond the infected point and beyond that a 5 km radius where all fowl was to be vaccinated. This culling method was indiscriminate as a large proportion of the poultry inside these areas were small backyard flocks which did not travel great enough distances to carry infection to adjacent villages without human effort and may have not been infected at all. Between 2004 and 2005, over 100 million chickens were culled in Asia to contain H5N1.

The risk of mass culling of birds and the resulting economic impact led farmers who were reluctant to report sick poultry. The culls often preempted actual lab testing for H5N1 as avian flu policy justified sacrificing poultry as a safeguard against HPAI spread. In response to these policies, farmers in Vietnam between 2003 and 2004 became more and more unwilling to surrender apparently healthy birds to authorities and stole poultry destined for culls as it stripped poultry of their biosocial and economic worth. By the end of 2005, the government implemented a new policy that targeted high-risk flock in the immediate vicinity of infected farms and instituted voluntary culling with compensation in the case of a local outbreak.

Not only did culling result in severe economic impacts especially for small scale farmers, culling itself may be an ineffective preventative measure. In the short-term, mass culling achieves it goals of limiting the immediate spread of HPAI, it has been found to impede the evolution of host resistance which is important for the long-term success of HPAI control. Mass culling also selects for elevated influenza virulence and results in the greater mortality of birds overall. Effective culling strategies must be selective as well as considerate of economic impacts to optimize epidemiological control and minimize economic and agricultural destruction.

People-poultry Relations

Prevention and control programs must take into account local understandings of people-poultry relations. In the past, programs that have focused on singular, place-based understandings of disease transmission have been ineffective. In the case of Northern Vietnam, health workers saw poultry as commodities with an environment that was under the control of people. Poultry existed in the context of farms, markets, slaughterhouses, and roads while humans were indirectly the primary transmitters of avian flu, placing the burden of disease control on people. However, farmers saw their free ranging poultry in an environment dominated by nonhuman forces that they could not exert control over. There were a host of nonhuman actors such as wild birds and weather patterns whose relationships with the poultry fostered the disease and absolved farmers of complete responsibility for disease control.

Attempts at singular, place-based controls sought to teach farmers to identify areas where their behavior could change without looking at poultry behaviors. Behavior recommendations by Vietnam's National Steering Committee for Avian Influenza Control and Prevention (NSCAI) were drawn from the FAO Principles of Biosecurity. These included restrictions from entering areas where poultry are kept by erecting barriers to segregate poultry from non-human contact, limits on human movement of poultry and poultry-related products ideally to transporters, and recommendations for farmers to wash hands and footwear before and after contact with poultry. Farmers, pointed to wind and environmental pollution as reasons poultry would get sick. NSCAI recommendations also would disrupt longstanding livestock production practices as gates impede sales by restricting assessment of birds by appearance and offend customers by limiting outside human contact. Instead of incorporating local knowledge into recommendations, cultural barriers were used as scapegoats for failed interventions. Prevention and control methods have been more effective when also considering the social, political, and ecological agents in play.

Rhabdoviruses

Rhabdo in Greek means 'rod-shaped'. The first description of rabies dates from the 23rd century BC in Mesopotamia. In Latin word " rabere" means to rage (In sanskrit word Rabhas means to do violence). During early 1881, French chemist Louis Pasteur and his assistant, Emile Roux , begin their research on rabies and successfully vaccinated Joseph Meister who was bitten by a rabid dog. Rabies is a zoonotic viral disease most often transmitted after the bite of rabies infected animal. Large cases of the rabies are reported each year from different parts of the world. The members of the family *Rhabdoviridae* belong to order *Mononegavirales* and contain six genera. The Rhabdoviruses contain negative strand RNA genome ranging from 11 to 15 Kb in size. Those Rhabdoviruses that infect plants are rod shaped with rounded ends eg Potato yellow dwarf virus while those infecting animals are bullet shaped eg Rabies virus (Lyssa virus).

Rhabdoviridae is divided into six genera

1. Novirhabdovirus	↑	
2. Vesiculovirus		Animal viruses
3. Lyssa virus		
4. Ephemerovirus	↓	

1. Cytorhabdovirus 2. Nucleorhabdovirus	↕	Plant viruses

Virus	Virus type	Host species
Vesiculovirus	Vesicular stomatitis Indiana virus	Vertebrates
Lyssavirus	Rabies virus	Vertebrates
Ephemerovirus	Bovine ephemeral fever virus	Vertebrates
Novirhabdovirus	Infectious haematopoetic necrosis virus	Vertebrates
Cytorhabdovirus	Lettuce necrotic yellows virus	Plant
Nucleorhabdovirus	Potato yellow dwarf virus	Plant

Virion Properties

Rhabdoviruses contain a linear, single stranded, negative sense RNA genome. Virions are 45- 100 nm in diameter and 100- 430 nm long. The virion has a cylindrical nucleocapsid surrounded by an envelope with large glycoprotein spikes. The virions are sensitive to heat and UV rays but stable towards the changes in pH. The virus is bullet shaped which is due to its lipid envelope. The genome is 11.9 kb in size which encodes for 5 genes in the following order.

Schematic representation of rhabdovirus virion:

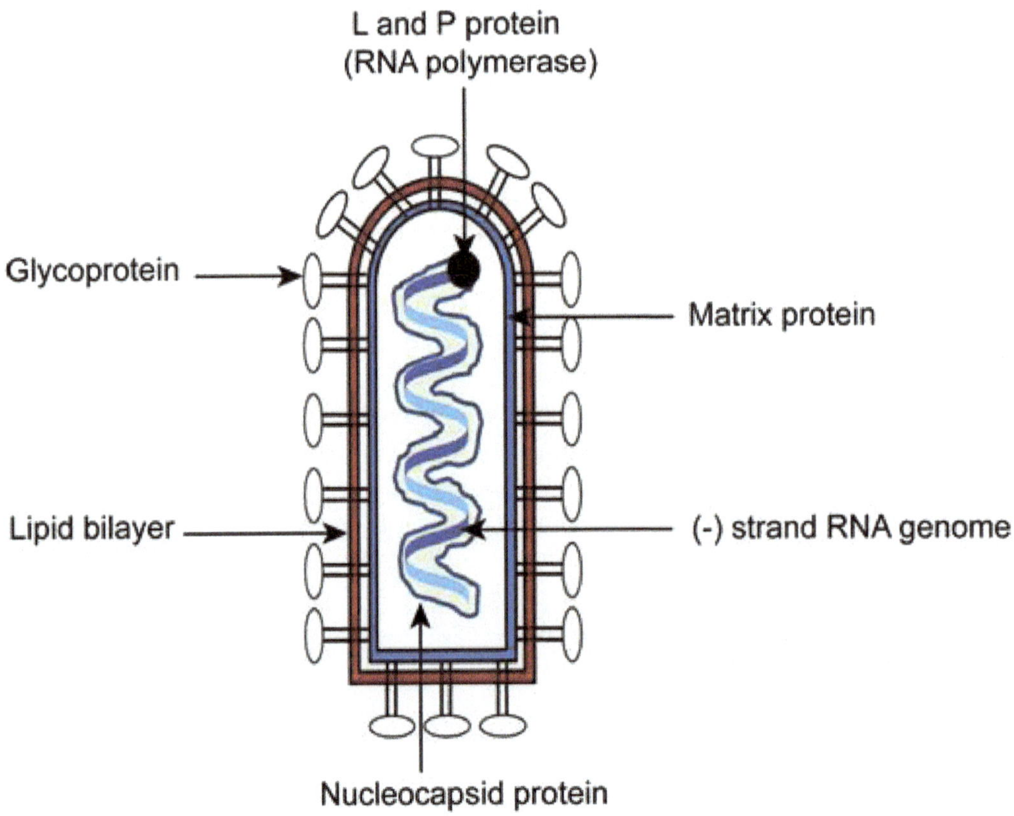

3' –N-P-M-G-L- 5'

N- Nucleocapsid protein

P- Phosphoprotein- cofactor of the viral polymerase

M- Inner virion protein/ helps in budding of the virion.

G- Glycoprotein that assists in making virion spikes

L- Large protein that represents RNA dependent RNA polymerase and helps in transcription and replication.

Virus Replication

In animals, virus replication takes place in the cytoplasm while in plants it may occur in nucleus. The virus entry into the host cell occurs by receptor mediated endocytosis followed by pH dependent fusion of virus envelope with endosomal membrane. As a result of fusion, the nucleocapsid is released into the cytoplasm. The first step of replication involves mRNA transcription from genomic RNA using RNA dependent RNA polymerase. For successful replication a large amount of nucleoprotein (N) and phosphoprotein (P) should be expressed. Switching of transcription to positive sense antigenome occurs after a threshold amount of N and P, which are then further used as a template for synthesis of negative stranded genomic RNA. There is a single promoter site at the 3' end of the viral genome where the polymerase attaches to the genomic RNA template and moves along the viral RNA. While moving it hits with start – stop signals at both the ends of the viral genes. Due to this only a small fraction undergoes continuous transcription process and hence this phenomenon is also known as attenuated transcription. Consequently more mRNA is produced towards the genes that are located at the 3' end and hence producing a gradient of mRNA in the order of N>P>M>G>L. As a result of the mRNA gradient, large amount of structural protein such as nucleocapsid protein is produced as compared to L protein.

Rabies

Rabies is a viral disease that causes inflammation of the brain in humans and other mammals. Early symptoms can include fever and tingling at the site of exposure. These symptoms are followed by one or more of the following symptoms: violent movements, uncontrolled excitement, fear of water, an inability to move parts of the body, confusion, and loss of consciousness. Once symptoms appear, the result is nearly always death. The time period between contracting the disease and the start of symptoms is usually one to three months; however, this time period can vary from less than one week to more than one year. The time is dependent on the distance the virus must travel to reach the central nervous system.

Rabies is caused by *lyssaviruses*, including rabies virus and Australian bat lyssavirus. Rabies is spread when an infected animal scratches or bites another animal or human. Saliva from an infected animal can also transmit rabies if the saliva comes into contact with the eyes, mouth, or nose. Globally, dogs are the most common animal involved. More than 99% of rabies cases in countries where dogs commonly have the disease are caused by dog bites. In the Americas, bat bites are the most common source of rabies infections in humans, and less than 5% of cases are from dogs. Rodents are very rarely infected with rabies. The rabies virus travels to the brain by following the peripheral nerves. The disease can only be diagnosed after the start of symptoms.

Animal control and vaccination programs have decreased the risk of rabies from dogs in a number of regions of the world. Immunizing people before they are exposed is recommended for those who are at high risk. The high-risk group includes people who work with bats or who spend prolonged periods in areas of the world where rabies is common. In people who have been exposed to rabies, the rabies vaccine and sometimes rabies immunoglobulin are effective in preventing the disease if the person receives the treatment before the start of rabies symptoms. Washing bites and scratches for 15 minutes with soap and water, povidone iodine, or detergent may reduce the number of viral particles and may be somewhat effective at preventing transmission. Only five people have survived a rabies infection after showing symptoms, and this was with extensive treatment known as the Milwaukee protocol.

Rabies causes about 24,000 to 60,000 deaths worldwide per year. More than 95% of human deaths caused by rabies occur in Africa and Asia. About 40% of deaths occur in children under the age of 15. Rabies is present in more than 150 countries and on all continents but Antarctica. More than 3 billion people live in regions of the world where rabies occurs. A number of countries, including Australia, Canada, Japan, and the United States, and Western Europe, do not have rabies among dogs. Many small island nations do not have rabies at all.

Signs and Symptoms

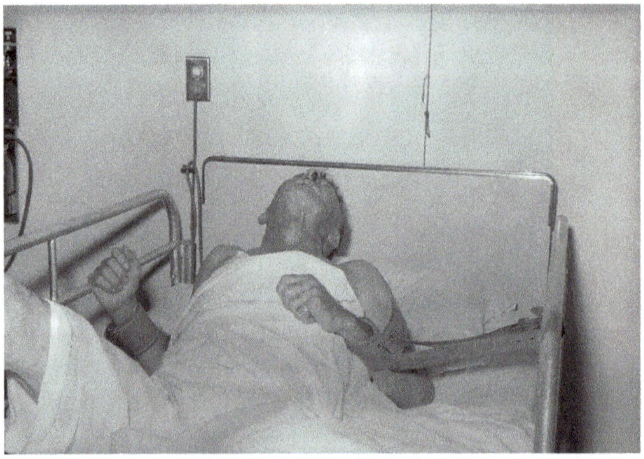

A person with rabies, 1959

The period between infection and the first symptoms (incubation period) is typically 1–3 months in humans. Incubation periods as short as four days and longer than six years have been documented, depending on the location and severity of the contaminated wound and the amount of virus introduced. Initial signs and symptoms of rabies are often nonspecific such as fever and headache. As rabies progresses and causes inflammation of the brain and/or meninges, signs and symptoms can include slight or partial paralysis, anxiety, insomnia, confusion, agitation, abnormal behavior, paranoia, terror, and hallucinations, progressing to delirium, and coma. The person may also have hydrophobia.

Death usually occurs 2 to 10 days after first symptoms. Survival is rare once symptoms have presented, even with the administration of proper and intensive care. Jeanna Giese, who in 2004 was the first patient treated with the Milwaukee protocol, became the first person ever recorded to have survived rabies without receiving successful post-exposure prophylaxis. An intention-to-treat analysis has since found this protocol has a survival rate of about 8%.

Hydrophobia

A rabid dog

Hydrophobia ("fear of water") is the historic name for rabies. It refers to a set of symptoms in the later stages of an infection in which the person has difficulty swallowing, shows panic when presented with liquids to drink, and cannot quench his or her thirst. Any mammal infected with the virus may demonstrate hydrophobia.

Saliva production is greatly increased, and attempts to drink, or even the intention or suggestion of drinking, may cause excruciatingly painful spasms of the muscles in the throat and larynx. This can be attributed to the fact that the virus multiplies and assimilates in the salivary glands of the infected animal for the purpose of further transmission through biting. The ability to transmit the virus would decrease significantly if the infected individual could swallow saliva and water.

Hydrophobia is commonly associated with furious rabies, which affects 80% of rabies-infected people. The remaining 20% may experience a paralytic form of rabies that is marked by muscle weakness, loss of sensation, and paralysis; this form of rabies does not usually cause fear of water.

Cause

Rabies is caused by a number of *lyssaviruses* including: rabies virus and Australian bat lyssavirus.

The rabies virus is the type species of the *Lyssavirus* genus, in the family *Rhabdoviridae*, order *Mononegavirales*. Lyssavirions have helical symmetry, with a length of about 180 nm and a cross-section of about 75 nm. These virions are enveloped and have a single-stranded RNA genome with negative sense. The genetic information is packed as a ribonucleoprotein complex in which RNA is tightly bound by the viral nucleoprotein. The RNA genome of the virus encodes five genes whose order is highly conserved: nucleoprotein (N), phosphoprotein (P), matrix protein (M), glycoprotein (G), and the viral RNA polymerase (L).

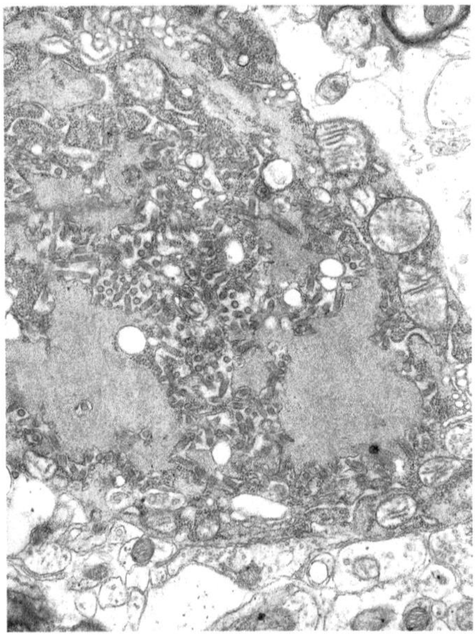

TEM micrograph with numerous rabies virions (small, dark grey, rodlike particles) and Negri bodies (the larger pathognomonic cellular inclusions of rabies infection)

Once within a muscle or nerve cell, the virus undergoes replication. The trimeric spikes on the exterior of the membrane of the virus interact with a specific cell receptor, the most likely one being the acetylcholine receptor. The cellular membrane pinches in a procession known as pinocytosis and allows entry of the virus into the cell by way of an endosome. The virus then uses the acidic environment, which is necessary, of that endosome and binds to its membrane simultaneously, releasing its five proteins and single strand RNA into the cytoplasm.

The L protein then transcribes five mRNA strands and a positive strand of RNA all from the original negative strand RNA using free nucleotides in the cytoplasm. These five mRNA strands are then translated into their corresponding proteins (P, L, N, G and M proteins) at free ribosomes in the cytoplasm. Some proteins require post-translative modifications. For example, the G protein travels through the rough endoplasmic reticulum, where it undergoes further folding, and is then transported to the Golgi apparatus, where a sugar group is added to it (glycosylation).

Where there are enough proteins, the viral polymerase will begin to synthesize new negative strands of RNA from the template of the positive strand RNA. These negative strands will then form complexes with the N, P, L and M proteins and then travel to the inner membrane of the cell, where a G protein has embedded itself in the membrane. The G protein then coils around the N-P-L-M complex of proteins taking some of the host cell membrane with it, which will form the new outer envelope of the virus particle. The virus then buds from the cell.

From the point of entry, the virus is neurotropic, traveling quickly along the neural pathways into the central nervous system. The virus usually first infects muscle cells close to the site of infection, where they are able to replicate without being 'noticed' by the host's immune system. Once enough virus has been replicated, they begin to bind to acetylcholine receptors (p75NR) at the neuromuscular junction. The virus then travels through the nerve cell axon via retrograde transport, as its

P protein interacts with dynein, a protein present in the cytoplasm of nerve cells. Once the virus reaches the cell body it travels rapidly to the Central Nervous System (CNS), replicating in motor neurons and eventually reaching the brain. After the brain is infected, the virus travels centrifugally to the peripheral and autonomic nervous systems, eventually migrating to the salivary glands, where it is ready to be transmitted to the next host.

Transmission

All warm-blooded species, including humans, may become infected with the rabies virus and develop symptoms. Birds were first artificially infected with rabies in 1884; however, infected birds are largely, if not wholly, asymptomatic, and recover. Other bird species have been known to develop rabies antibodies, a sign of infection, after feeding on rabies-infected mammals.

The virus has also adapted to grow in cells of cold-blooded vertebrates. Most animals can be infected by the virus and can transmit the disease to humans. Infected bats, monkeys, raccoons, foxes, skunks, cattle, wolves, coyotes, dogs, mongooses (normally yellow mongoose) and cats present the greatest risk to humans.

Rabies may also spread through exposure to infected bears, domestic farm animals, groundhogs, weasels, and other wild carnivorans. Lagomorphs, such as hares and rabbits, and small rodents such as chipmunks, gerbils, guinea pigs, hamsters, mice, rats, and squirrels, are almost never found to be infected with rabies and are not known to transmit rabies to humans. Bites from mice, rats, or squirrels rarely require rabies prevention because these rodents are typically killed by any encounter with a larger, rabid animal, and would, therefore, not be carriers. The Virginia opossum is resistant but not immune to rabies.

The virus is usually present in the nerves and saliva of a symptomatic rabid animal. The route of infection is usually, but not always, by a bite. In many cases, the infected animal is exceptionally aggressive, may attack without provocation, and exhibits otherwise uncharacteristic behavior. This is an example of a viral pathogen modifying the behavior of its host to facilitate its transmission to other hosts.

Transmission between humans is extremely rare. A few cases have been recorded through transplant surgery. The only well-documented cases of rabies caused by human-to-human transmission occurred among eight recipients of transplanted corneas and among three recipients of solid organs. In addition to transmission from cornea and organ transplants, bite and non-bite exposures inflicted by infected humans could theoretically transmit rabies, but no such cases have been documented, since infected humans are usually hospitalized and necessary precautions taken. Casual contact, such as touching a person with rabies or contact with non-infectious fluid or tissue (urine, blood, feces) does not constitute an exposure and does not require post-exposure prophylaxis. Additionally, as the virus is present in sperm or vaginal secretions, spread through sex may be possible.

After a typical human infection by bite, the virus enters the peripheral nervous system. It then travels along the afferent nerves toward the central nervous system. During this phase, the virus cannot be easily detected within the host, and vaccination may still confer cell-mediated immunity to prevent symptomatic rabies. When the virus reaches the brain, it rapidly causes encephalitis, the prodromal phase, which is the beginning of the symptoms. Once the patient becomes symptomatic, treatment is almost never effective and mortality is over 99%. Rabies may also inflame the spinal cord, producing transverse myelitis.

Diagnosis

Rabies can be difficult to diagnose, because, in the early stages, it is easily confused with other diseases or with aggressiveness. The reference method for diagnosing rabies is the fluorescent antibody test (FAT), an immunohistochemistry procedure, which is recommended by the World Health Organization (WHO). The FAT relies on the ability of a detector molecule (usually fluorescein isothiocyanate) coupled with a rabies-specific antibody, forming a conjugate, to bind to and allow the visualisation of rabies antigen using fluorescent microscopy techniques. Microscopic analysis of samples is the only direct method that allows for the identification of rabies virus-specific antigen in a short time and at a reduced cost, irrespective of geographical origin and status of the host. It has to be regarded as the first step in diagnostic procedures for all laboratories. Autolysed samples can, however, reduce the sensitivity and specificity of the FAT. The RT PCR assays proved to be a sensitive and specific tool for routine diagnostic purposes, particularly in decomposed samples or archival specimens. The diagnosis can be reliably made from brain samples taken after death. The diagnosis can also be made from saliva, urine, and cerebrospinal fluid samples, but this is not as sensitive and reliable as brain samples. Cerebral inclusion bodies called Negri bodies are 100% diagnostic for rabies infection but are found in only about 80% of cases. If possible, the animal from which the bite was received should also be examined for rabies.

Some light microscopy techniques may also be used to diagnose rabies at a tenth of the cost of traditional fluorescence microscopy techniques, allowing identification of the disease in less-developed countries.

Differential Diagnosis

The differential diagnosis in a case of suspected human rabies may initially include any cause of encephalitis, in particular infection with viruses such as herpesviruses, enteroviruses, and arboviruses such as West Nile virus. The most important viruses to rule out are herpes simplex virus type one, varicella zoster virus, and (less commonly) enteroviruses, including coxsackieviruses, echoviruses, polioviruses, and human enteroviruses 68 to 71.

New causes of viral encephalitis are also possible, as was evidenced by the 1999 outbreak in Malaysia of 300 cases of encephalitis with a mortality rate of 40% caused by Nipah virus, a newly recognized paramyxovirus. Likewise, well-known viruses may be introduced into new locales, as is illustrated by the outbreak of encephalitis due to West Nile virus in the eastern United States. Epidemiologic factors, such as season, geographic location, and the patient's age, travel history, and possible exposure to bites, rodents, and ticks, may help direct the diagnosis.

Prevention

Almost all human cases of rabies were fatal until a vaccine was developed in 1885 by Louis Pasteur and Émile Roux. Their original vaccine was harvested from infected rabbits, from which the virus in the nerve tissue was weakened by allowing it to dry for five to ten days. Similar nerve tissue-derived vaccines are still used in some countries, as they are much cheaper than modern cell culture vaccines.

The human diploid cell rabies vaccine was started in 1967. Less expensive purified chicken embryo cell vaccine and purified vero cell rabies vaccine are now available. A recombinant vaccine called V-RG has been used in Belgium, France, Germany, and the United States to prevent outbreaks of rabies in undomesticated animals. Immunization before exposure has been used in both human and nonhuman populations, where, as in many jurisdictions, domesticated animals are required to be vaccinated.

The number of recorded human deaths from rabies in the United States has dropped from 100 or more annually in the early 20th century to one or two per year due to widespread vaccination of domestic dogs and cats and the development of human vaccines and immunoglobulin treatments. Most deaths now result from bat bites, which may go unnoticed by the victim and hence untreated.

The Missouri Department of Health and Senior Services Communicable Disease Surveillance 2007 Annual Report states the following can help reduce the risk of contracting rabies:

- Vaccinating dogs, cats, and ferrets against rabies
- Keeping pets under supervision
- Not handling wild animals or strays
- Contacting an animal control officer upon observing a wild animal or a stray, especially if the animal is acting strangely
- If bitten by an animal, washing the wound with soap and water for 10 to 15 minutes and contacting a healthcare provider to determine if post-exposure prophylaxis is required

September 28 is World Rabies Day, which promotes the information, prevention, and elimination of the disease.

Treatment

Treatment after exposure can prevent the disease if administered promptly, generally within 10 days of infection. Thoroughly washing the wound as soon as possible with soap and water for approximately five minutes is effective in reducing the number of viral particles. Povidone-iodine or alcohol is then recommended to reduce the virus further.

In the US, the Centers for Disease Control and Prevention recommends people receive one dose of human rabies immunoglobulin (HRIG) and four doses of rabies vaccine over a 14-day period. The immunoglobulin dose should not exceed 20 units per kilogram body weight. HRIG is expensive and constitutes most of the cost of post exposure treatment, ranging as high as several thousand dollars. As much as possible of this dose should be injected around the bites, with the remainder being given by deep intramuscular injection at a site distant from the vaccination site.

The first dose of rabies vaccine is given as soon as possible after exposure, with additional doses on days three, seven and 14 after the first. Patients who have previously received pre-exposure vaccination do not receive the immunoglobulin, only the postexposure vaccinations on days 0 and 3.

The pain and side effects of modern cell-based vaccines are similar to flu shots. The old nerve-tissue-based vaccinations that require multiple painful injections into the abdomen with a large needle are inexpensive, but are being phased out and replaced by affordable World Health Organization intradermal-vaccination regimens.

Intramuscular vaccination should be given into the deltoid, not the gluteal area, which has been associated with vaccination failure due to injection into fat rather than muscle. In infants, the lateral thigh is recommended.

Awakening to find a bat in the room, or finding a bat in the room of a previously unattended child or mentally disabled or intoxicated person, is regarded as an indication for post-exposure prophylaxis (PEP). The recommendation for the precautionary use of PEP in occult bat encounters where no contact is recognized has been questioned in the medical literature, based on a cost–benefit analysis. However, a 2002 study has supported the protocol of precautionary administering of PEP where a child or mentally compromised individual has been alone with a bat, especially in sleep areas, where a bite or exposure may occur without the victim being aware. Begun with little or no delay, PEP is 100% effective against rabies. In the case in which there has been a significant delay in administering PEP, the treatment should be administered regardless, as it may still be effective. More than 15 million people get vaccination after potential exposure a year. While this works well the cost is significant.

Induced Coma

In 2004, American teenager Jeanna Giese survived an infection of rabies unvaccinated. She was placed into an induced coma upon onset of symptoms and given ketamine, midazolam, ribavirin, and amantadine. Her doctors administered treatment based on the hypothesis that detrimental effects of rabies were caused by temporary dysfunctions in the brain and could be avoided by inducing a temporary partial halt in brain function that would protect the brain from damage while giving the immune system time to defeat the virus. After 31 days of isolation and 76 days of hospitalization, Giese was released from the hospital. She survived with all higher level brain functions, but an inability to walk and balance. On a podcast of NPR's *Radiolab*, Giese recounted: "I had to learn how to stand and then to walk, turn around, move my toes. I was really, after rabies, a new born baby who couldn't do anything. I had to relearn that all...mentally I knew how to do stuff but my body wouldn't cooperate with what I wanted it to do. It definitely took a toll on me psychologically. You know I'm still recovering. I'm not completely back. Stuff like balance, and I can't run normally."

Giese's treatment regimen became known as the Milwaukee protocol, which has since undergone revision with the second version omitting the use of ribavirin. Two of 25 patients survived when treated under the first protocol. A further 10 patients have been treated under the revised protocol, with a further two survivors.

On June 12, 2011, Precious Reynolds, an eight-year-old girl from Humboldt County, California, became the third reported person in the United States to have recovered from rabies without receiving PEP.

Prognosis

In unvaccinated humans, rabies is almost always fatal after neurological symptoms have developed.

Vaccination after exposure, PEP, is highly successful in preventing the disease if administered promptly, in general within 6 days of infection. Begun with little or no delay, PEP is 100% effective against rabies. In the case of significant delay in administering PEP, the treatment still has a chance of success.

Five of the first 43 patients (12%) treated with the Milwaukee protocol survived, and those receiving treatment survived longer than those not receiving the treatment.

Epidemiology

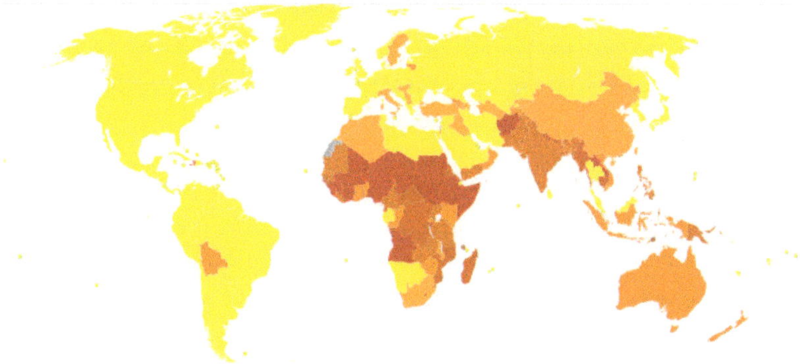

Deaths from rabies per million persons in 2012

	0
	1
	2–4
	5–9
	10–17
	18–69

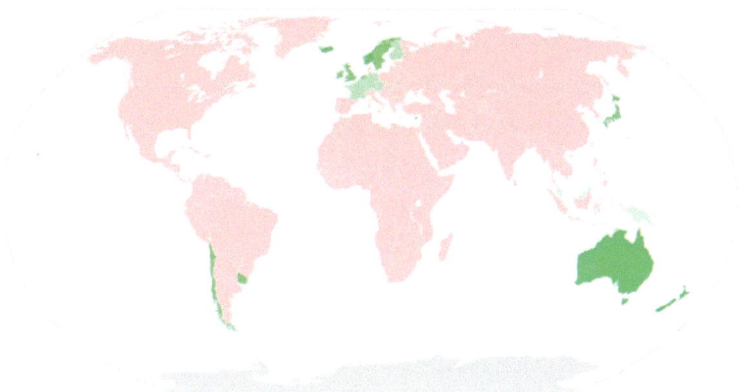

Rabies-free countries (in green) as of 2013. always rabies-free
rabies eliminated before 1990
rabies eliminated in or after 1990
year of rabies elimination unknown

In 2010, an estimated 26,000 people died from rabies, down from 54,000 in 1990. The majority of the deaths occurred in Asia and Africa. India has the highest rate of human rabies in the world, primarily because of stray dogs, whose number has greatly increased since a 2001 law forbade the killing of dogs. Effective control and treatment of rabies in India is also hindered by a form of mass hysteria known as puppy pregnancy syndrome (PPS). Dog bite victims with PPS (both male and female) become convinced that puppies are growing inside them, and often seek help from faith healers rather than from conventional medical services. In cases where the bite was from a rabid dog, this decision can prove fatal. Dr. Nitai Kishore Marik, former district medical officer of West Midnapur, states "I have seen scores of cases of rabies that reached our hospitals very late because of the intervention of faith healers. We could not save those lives." An estimated 20,000 people die every year from rabies in India — more than a third of the global toll. As of 2015, China had the second-highest number of cases (approximately 6,000), followed by the Democratic Republic of the Congo (5,600).

The rabies virus survives in widespread, varied, rural animal reservoirs. Despite Australia's official rabies-free status, Australian bat lyssavirus (ABLV), discovered in 1996, is a strain of rabies prevalent in native bat populations. There have been three human cases of ABLV in Australia, all of them fatal.

In Asia and in parts of the Americas and Africa, dogs remain the principal host. Mandatory vaccination of animals is less effective in rural areas. Especially in developing countries, pets may not be privately kept and their destruction may be unacceptable. Oral vaccines can be safely distributed in baits, a practice that has successfully reduced rabies in rural areas of Canada, France, and the United States. In Montréal, Quebec, Canada, baits are successfully used on raccoons in the Mont-Royal Park area. Vaccination campaigns may be expensive, and cost-benefit analysis suggests baits may be a cost-effective method of control. In Ontario, a dramatic drop in rabies was recorded when an aerial bait-vaccination campaign was launched.

Rabies is common among wild animals in the US. Bats, raccoons, skunks and foxes account for almost all reported cases (98% in 2009). Rabid bats are found in all 48 contiguous states. Other reservoirs are more limited geographically; for example, the raccoon rabies virus variant is only found in a relatively narrow band along the East Coast. Due to a high public awareness of the virus, efforts at vaccination of domestic animals and curtailment of feral populations, and availability of postexposure prophylaxis, incidents of rabies in humans are very rare. A total of 49 cases of the disease was reported in the country between 1995 and 2011; of these, 11 are thought to have been acquired abroad. Almost all domestically acquired cases are attributed to bat bites.

In Switzerland, the disease has been virtually eliminated after scientists placed chicken heads laced with live attenuated vaccine in the Swiss Alps. The foxes of Switzerland, proven to be the main source of rabies in the country, ate the chicken heads and immunized themselves.

Italy, after being declared rabies-free from 1997 to 2008, has witnessed a reemergence of the disease in wild animals in the Triveneto regions (Trentino-Alto Adige/Südtirol, Veneto and Friuli-Venezia Giulia), due to the spreading of an epidemic in the Balkans that hit Austria too. An extensive wild animals vaccination campaign eliminated the virus from Italy again, and it regained the rabies-free country status in 2013, the last reported case of rabies being reported in a red fox in early 2011.

A 2015 collaboration between the World Health Organization, World Organization of Animal Health (OIE), Food and Agriculture Organization of the United Nation (FAO), and Global Alliance for Rabies Control has a goal of eliminating deaths from rabies by 2030.

History

A woodcut from the Middle Ages showing a rabid dog.

Rabies has been known since around 2000 B.C. The first written record of rabies is in the Mesopotamian Codex of Eshnunna (circa 1930 BC), which dictates that the owner of a dog showing symptoms of rabies should take preventive measure against bites. If another person were bitten by a rabid dog and later died, the owner was heavily fined.

Ineffective folk remedies abounded in the medical literature of the ancient world. The physician Scribonius Largus prescribed a poultice of cloth and hyena skin; Antaeus recommended a preparation made from the skull of a hanged man.

Rabies appears to have originated in the Old World, the first epizootic in the New World occurring in Boston in 1768. It spread from there, over the next few years, to various other states, as well as to the French West Indies, eventually becoming common all across North America.

Rabies was considered a scourge for its prevalence in the 19th century. In France and Belgium, where Saint Hubert was venerated, the "St Hubert's Key" was heated and applied to cauterize the wound. By an application of magical thinking, dogs were branded with the key in hopes of protecting them from rabies. The fear of rabies was almost irrational, due to the insignificant number of vectors (mostly rabid dogs) and the absence of any efficacious treatment. It was not uncommon for a person bitten by a dog but merely suspected of being rabid, to commit suicide or to be killed by others. This gave Louis Pasteur ample opportunity to test postexposure treatments from 1885. In ancient times, the attachment of the tongue (the lingual frenulum, a mucous membrane) was cut and removed as this is where rabies was thought to originate. This practice ceased with the discovery of the actual cause of rabies.

In modern times, the fear of rabies has not diminished, and the disease and its symptoms, particularly agitation has served as an inspiration for several works of zombie or similarly-themed fiction, often portraying rabies as having mutated into a stronger virus which fills humans with murderous rage or uncurable illness, bringing about a devastating, widespread pandemic.

Etymology

The term is derived from the Latin *rabies*, "madness". This, in turn, may be related to the Sanskrit *rabhas*, "to do violence". The Greeks derived the word *lyssa*, from *lud* or "violent"; this root is used in the genus name of the rabies virus, *Lyssavirus*.

Other Animals

Rabies is infectious to mammals; three stages of central nervous system infection are recognized. The first stage is a one- to three-day period characterized by behavioral changes and is known as the prodromal stage. The second is the excitative stage, which lasts three to four days. This stage is often known as "furious rabies" for the tendency of the affected animal to be hyper-reactive to external stimuli and bite at anything near. The third is the paralytic stage and is caused by damage to motor neurons. Incoordination is seen, owing to rear limb paralysis, and drooling and difficulty swallowing is caused by paralysis of facial and throat muscles. Death is usually caused by respiratory arrest.

Research

Rabies has the advantage over other pseudotyping methods for gene delivery in that the cell-targeting (tissue tropism) is more specific for difficult-to-reach sites, such as the central nervous system without invasive delivery methods, as well as being capable of retrograde tracing (i.e., going against the flow of information at synapses) in neuronal circuits.

Evidence indicates artificially increasing the permeability of the blood–brain barrier, which normally does not allow most immune cells across, promotes viral clearance.

Pathogenesis and Pathology

In rabies the prognosis of the disease is mainly decided by the location and severity of the site of bite and the species of animals involved. Rabies has three stages. The first stage is characterized by behavioral changes also known as prodromal stage. In the second stage called as excitative stage the animal exhibits disease in furious forms and in the third stage the animal manifests the paralytic form of the disease. After entry the virus first affects the peripheral nervous system and replicates in the brain. Dumb and paralytic forms of the disease appear as it progresses towards the central nervous system. In the nervous system the virus is formed by budding in various membranes and glands. Neurons accumulate ribonucleoprotein as intracytoplasmic inclusion bodies often termed as Negri bodies and are pathognomonic for rabies. Salivary glands help the virus to bud on plasma membrane and release it in very high concentrations through the saliva. Death usually occurs due to respiratory arrest .

Progression of rabies virus through neurons in body:

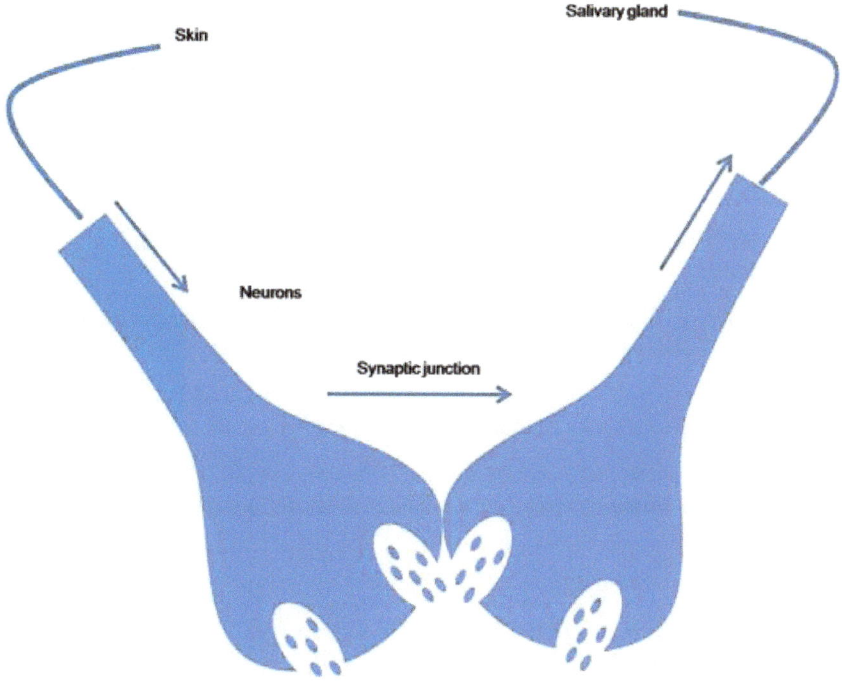

Diagnosis

Any patient having a history of animal bite should be suspected for rabies. Differential diagnosis with West Nile virus, herpesvirus, enterovirus is required. History and clinical symptoms like encephalitis, salivation, etc provide concrete support for rabies. Besides this, Negri bodies in the neurons of affected animals can be visualized using Seller's stain. RT-PCR assay can be conducted to test for the presence of viral RNA in the brain of suspected animal. RT-PCR of saliva can also be carried out. Immuno-histochemical staining can also be done from frozen sections of brain tissue. Fluorescent antibody test can also be done for the detection of viral antigens from the tissue samples.

Immunity, Prevention and Control

Rabies is a highly immunogenic disease and many vaccines are available for the same. In early stages of infection, infectious rabies virus is susceptible to antibody mediated neutralization and this efficacy has been proved in exposed humans of the classical Pasteurian post- exposure vaccination and it provides even better results when administered along with hyper-immune globulin.

Blood brain barrier shows promising results when its permeability is increased artificially as it helps in viral clearance and does not allow most immune cells across. Thus altering the permeability of blood brain barrier to contribute to viral clearance can also be exploited in case of Rabies.

Besides this, prophylactic measures should be taken like scheduled vaccination, washing the wounds with soap for 5 to 10 times while treating the infected cases and regular spaying of the animals can prevent the disease.

References

- "UC Davis Children's Hospital patient becomes third person in U.S. to survive rabies". UC Davis Medical Center. Archived from the original on 21 May 2012. Retrieved 3 May 2012
- "Sequence Comparison of Five Polymerases (L proteins) of Unsegmented Negative-strand RNA Viruses: Theoretical Assignment of Functional Domains". Journal of General Virology. 71
- Klein, Donald W.; Prescott, Lansing M.; Harley, John (1993). Microbiology. Dubuque, Iowa: Wm. C. Brown. ISBN 0-697-01372-3
- Rudi Hendrikx, DVM. "Avian Influenza in Pigeons" (PDF). Archived from the original (PDF) on 2014-02-22. Retrieved 2014-02-06
- "Replication- competent rhabdoviruses with human immunodeficiency virus type 1 coats and green fluorescent protein: entry by a pH-independent pathway". Journal of Virology. 73
- Cox, N.; Kawaoka (1998). "22". In Mahy B. and Collier L. Topley and Wilson's Microbiology and Microbial Infections. 1 Virology. Y. (9 ed.). Arnold. p. 415. ISBN 0-340-61470-6
- Leong HK, Goh CS, Chew ST, et al. (June 2008). "Prevention and control of avian influenza in Singapore" (PDF). Ann. Acad. Med. Singap. 37 (6): 504–9. PMID 18618063. Retrieved 2009-04-15
- "Sendai Virus and Simian Virus 5 Block Activation of Interferon-Responsive Genes: Importance for Virus Pathogenesis". Journal of Virology. 73
- Cotran RS, Kumar V, Fausto N (2005). Robbins and Cotran Pathologic Basis of Disease (7th ed.). Elsevier/ Saunders. p. 1375. ISBN 0-7216-0187-1
- "Cumulative number of confirmed human cases for avian influenza A(H5N1) reported to WHO, 2003-2012" (PDF). World Health Organization. Retrieved 5 September 2012
- Hiromoto, Yasuaki. "Evolutionary characterization of the six internal genes of H5N1 human influenza A virus". Journal of General Virology. Retrieved 2013-04-20
- Tintinalli, Judith E. (2010). Emergency Medicine: A Comprehensive Study Guide (Emergency Medicine (Tintinalli)). McGraw-Hill. pp. Chapter 152. ISBN 0-07-148480-9
- "The fight against bird flu". Nature. 496 (7446): 397. April 24, 2013. PMID 23627002. doi:10.1038/496397a. Retrieved 25 April 2013
- William H. Wunner (2010). Rabies: Scientific Basis of the Disease and Its Management. Academic Press. p. 556. ISBN 9780080550091
- "Frequently Asked Questions on human infection with influenza A(H7N9) virus, China". World Health Organization. 5 April 2013. Retrieved 9 April 2013
- Shim, E.; Galvani, A. P. (2009). "Evolutionary repercussions of avian culling on host resistance and influenza virulence". PLoS ONE. 4 (5): e5503. doi:10.1371/journal.pone.0005503
- WHO Expert Consultation on Rabies : second report. (PDF) (2 ed.). Geneva: WHO. 2013. p. 3. ISBN 9789241209823
- Gough PM, Jorgenson RD (1976). "Rabies antibodies in sera of wild birds". Journal of Wildlife Diseases. 12 (3): 392–5. doi:10.7589/0090-3558-12.3.392. PMID 16498885
- Smallman-Raynor, Andrew Cliff, Peter Haggett, Matthew (2004). World atlas of epidemic diseases. London: Arnold. p. 51. ISBN 9780340761717
- "OIE expert mission finds live bird markets play a key role in poultry and human infections with influenza A(H7N9)". Paris: World Organisation for Animal Health. April 30, 2013. Retrieved May 2, 2013

5

Diverse Groups of RNA Viruses

A major known RNA virus is double-stranded RNA viruses, which can acquire and survive in a range of hosts including humans, animals, fungi, plants etcetera. Of this, Reoviridae is the most diverse virus. These reoviruses can cause varied number of diseases including respiratory and enteric diseases, Colorado tick fever, gastroenteritis and diarrhoea. The chapter serves as a source to understand the major categories related to the varied types of RNA viruses.

Double-stranded RNA Viruses

Double-stranded (ds) RNA viruses are a diverse group of viruses that vary widely in host range (humans, animals, plants, fungi, and bacteria), genome segment number (one to twelve) and virion organization (T-number, capsid layers or turrets). Members of this group include the rotaviruses, known globally as a common cause of gastroenteritis in young children, and bluetongue virus, an economically important pathogen of cattle and sheep.

Of these families, the *Reoviridae* is the largest and most diverse in terms of host range.

In recent years the increasing knowledge of virus particle assembly, virus-cell interactions, and viral pathogenesis allow approaches for the development of novel antiviral strategies or agents.

Taxonomy

Viruses with dsRNA genomes are currently grouped into a number of families, unassigned genera and species.

Three families infect fungi: *Totiviridae, Partitiviridae* and *Chrysoviridae.* These families have monopartite, bipartite and quadripartite genomes respectively. They are typically isometric particles 25–50 nanometers in diameter. Based on sequence similarity of the RNA dependent RNA polymerase, the partitiviruses are probably derived from a totivirus ancestor. A fourth family — *Alternaviridae* — has recently been described also with quadripartite genome.

Hypoviruses are mycoviruses (fungal viruses) with unencapsidated dsRNA genomes. They may have common ancestry with plant positive strand RNA viruses in supergroup 1 with potyvirus lineages, respectively

A new clade (as yet unnamed) of six viruses infecting filamentous fungi has been reported.

Taxa

Families

- *Amalgaviridae*

- *Birnaviridae*
- *Chrysoviridae*
- *Cystoviridae*
- *Endornaviridae*
- *Hypoviridae*
- *Megabirnaviridae*
- *Partitiviridae*
- *Picobirnaviridae*
- *Quadriviridae*
- *Reoviridae*
- *Totiviridae*

Unassigned species

- *La France isometric virus*

Notes on Selected Species

Reoviridae

Reoviridae are currently classified into nine genera. The genomes of these viruses consist of 10 to 12 segments of dsRNA, each generally encoding one protein. The mature virions are non-enveloped. Their capsids, formed by multiple proteins, have icosahedral symmetry and are arranged generally in concentric layers. A distinguishing feature of the dsRNA viruses, irrespective of their family association, is their ability to carry out transcription of the dsRNA segments, under appropriate conditions, within the capsid. In all these viruses, the enzymes required for endogenous transcription are thus part of the virion structure.

Orthoreoviruses

The orthoreoviruses (reoviruses) are the prototypic members of the virus *Reoviridae* family and representative of the turreted members, which comprise about half the genera. Like other members of the family, the reoviruses are non-enveloped and characterized by concentric capsid shells that encapsidate a segmented dsRNA genome. In particular, reovirus has eight structural proteins and ten segments of dsRNA. A series of uncoating steps and conformational changes accompany cell entry and replication. High-resolution structures are known for almost all of the proteins of mammalian reovirus (MRV), which is the best-studied genotype. Electron cryo-microscopy (cryo-EM) and X-ray crystallography have provided a wealth of structural information about two specific MRV strains, type 1 Lang (T1L) and type 3 Dearing (T3D).

Cypovirus

The cytoplasmic polyhedrosis viruses (CPVs) form the genus Cypovirus of the family *Reoviridae*. CPVs are classified into 14 species based on the electrophoretic migration profiles of their genome segments. Cypovirus has only a single capsid shell, which is similar to the orthoreovirus inner core. CPV exhibits striking capsid stability and is fully capable of endogenous RNA transcription and processing. The overall folds of CPV proteins are similar to those of other reoviruses. However, CPV proteins have insertional domains and unique structures that contribute to their extensive intermolecular interactions. The CPV turret protein contains two methylase domains with a highly conserved helix-pair/β-sheet/helix-pair sandwich fold but lacks the β-barrel flap present in orthoreovirus λ2. The stacking of turret protein functional domains and the presence of constrictions and A spikes along the mRNA release pathway indicate a mechanism that uses pores and channels to regulate the highly coordinated steps of RNA transcription, processing, and release.

Rotavirus

Rotavirus is the most common cause of acute gastroenteritis in infants and young children worldwide. This virus contains a dsRNA genome and is a member of the *Reoviridae* family. The genome of rotavirus consists of eleven segments of dsRNA. Each genome segment codes for one protein with the exception of segment 11, which codes for two proteins. Among the twelve proteins, six are structural and six are non-structural proteins. It is a double-stranded RNA non-enveloped virus

Bluetongue Virus

The members of genus *Orbivirus* within the *Reoviridae* family are arthropod borne viruses and are responsible for high morbidity and mortality in ruminants. Bluetongue virus (BTV) which causes disease in livestock (sheep, goat, cattle) has been in the forefront of molecular studies for the last three decades and now represents the best understood orbivirus at the molecular and structural levels. BTV, like other members of the family, is a complex non-enveloped virus with seven structural proteins and a RNA genome consisting of 10 variously sized dsRNA segments.

Phytoreoviruses

Phytoreoviruses are non-turreted reoviruses that are major agricultural pathogens, particularly in Asia. One member of this family, Rice Dwarf Virus (RDV), has been extensively studied by electron cryomicroscopy and x-ray crystallography. From these analyses, atomic models of the capsid proteins and a plausible model for capsid assembly have been derived. While the structural proteins of RDV share no sequence similarity to other proteins, their folds and the overall capsid structure are similar to those of other *Reoviridae*.

The Yeast dsRNA Virus L-A

The L-A dsRNA virus of the yeast *Saccharomyces cerevisiae* has a single 4.6 kb genomic segment that encodes its major coat protein, Gag (76 kDa) and a Gag-Pol fusion protein (180 kDa) formed by a -1 ribosomal frameshift. L-A can support the replication and encapsidation in separate viral particles of any of several satellite dsRNAs, called M dsRNAs, each of which encodes a secreted protein toxin (the killer toxin) and immunity to that toxin. L-A and M are transmitted from cell to cell by the cytoplasmic mixing that occurs in the process of mating. Neither is naturally released from the cell or enters cells by other mechanisms, but the high frequency of yeast mating in nature results in the wide distribution of these viruses in natural isolates. Moreover, the structural and functional similarities with dsRNA viruses of mammals has made it useful to consider these entities as viruses.

Infectious Bursal Disease Virus

Infectious bursal disease virus (IBDV) is the best-characterized member of the family *Birnaviridae.* These viruses have bipartite dsRNA genomes enclosed in single layered icosahedral capsids with $T = 13l$ geometry. IBDV shares functional strategies and structural features with many other icosahedral dsRNA viruses, except that it lacks the $T = 1$ (or pseudo $T = 2$) core common to the *Reoviridae, Cystoviridae,* and *Totiviridae.* The IBDV capsid protein exhibits structural domains that show homology to those of the capsid proteins of some positive-sense single-stranded RNA viruses, such as the nodaviruses and tetraviruses, as well as the $T = 13$ capsid shell protein of the *Reoviridae.* The $T = 13$ shell of the IBDV capsid is formed by trimers of VP2, a protein generated by removal of the C-terminal domain from its precursor, pVP2. The trimming of pVP2 is performed on immature particles as part of the maturation process. The other major structural protein, VP3, is a multifunctional component lying under the $T = 13$ shell that influences the inherent structural polymorphism of pVP2. The virus-encoded RNA-dependent RNA polymerase, VP1, is incorporated into the capsid through its association with VP3. VP3 also interacts extensively with the viral dsRNA genome.

dsRNA Bacteriophage Φ6

Bacteriophage Φ6, is a member of the *Cystoviridae* family. It infects *Pseudomonas* bacteria (typically plant-pathogenic *P. syringae*). It has a three-part, segmented, double-stranded RNA genome, totalling ~13.5 kb in length. Φ6 and its relatives have a lipid membrane around their nucleocapsid, a rare trait among bacteriophages. It is a lytic phage, though under certain circumstances has been observed to display a delay in lysis which may be described as a "carrier state".

Anti-virals

Since cells do not produce double-stranded RNA during normal nucleic acid metabolism, natural selection has favored the evolution of enzymes that destroy dsRNA on contact. The best known class of this type of enzymes is Dicer. It is hoped that broad-spectrum anti-virals could be synthesized that take advantage of this vulnerability of double-stranded RNA viruses.

Total of 8 families of dsRNA viruses are currently recognized by ICTV. Two most important family members are Reoviruses and Birnaviruses. These two groups contain viruses of medical as well as of veterinary importance. Rotavirus is a major cause of infant diarrhea while blue tongue virus is a big issue for cattle and sheep. Infectious bursal disease virus is another important member of family *Birnaviridae* which causes an immunosuppressive disease in poultry and are of great economic importance. Another member of family *Birnaviridae* , Infectious pancreatic necrosis virus causes significant losses to the fisheries industry around the world.

Expression of Viral Proteins

The size of dsRNA genome of the viruses is limited by the icosahedral capsid which can accommodate the segmented genome and at the same time actively allow the transcription of the viral messages. The viruses evolve in such a way to conveniently dissociate the total translation products from the genome into several distinct proteins. Sometimes individual genome segments can also be exchanged in the presence of a suitable donor or acceptor which increases the genetic diversity in virus population. The dsRNA transcribes separately to full-length +ve sense RNA. The full-length +ve sense RNA acts as mRNA for viral protein synthesis as well as template for the synthesis of genomic RNA for the progeny virions.

Replication of the Genome

The replication of the genome is divided into several steps as follows:

1. In the first step primary transcription of the viral genome takes place inside viral core in the cytoplasm using viral RNA dependent RNA polymerase (RDRP)
2. Positive sense RNA is then transported into the cytoplasm
3. Positive sense RNA is translated to form viral proteins
4. Positive sense RNA and viral proteins are assembled to form immature virions
5. Positive sense RNA is then transcribed to form dsRNA in virions by viral RDRP
6. dsRNA undergoes secondary transcription
7. Final assembly and maturation of virions

Replication strategy of dsRNA virus

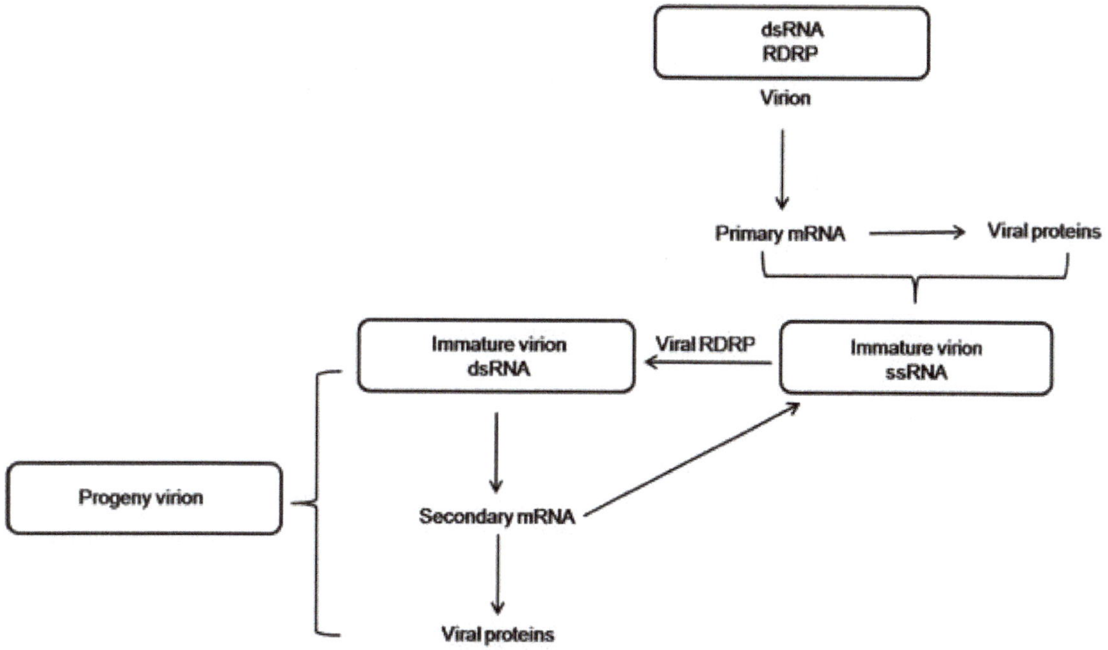

Reoviridae

Reoviridae is a family of viruses. They have a wide host range, including vertebrates, invertebrates, plants, and fungi. They are unique in that they lack lipid envelopes and package their genomes of discrete double-stranded segments of RNA within multi-layered capsids. Lack of a lipid envelope has allowed three-dimensional structures of these large complex viruses (diameter,~600–1,000 nm) to be obtained. There are currently 87 species in this family, divided among 30 genera. Reoviruses can affect the gastrointestinal system (such as *Rotavirus*) and respiratory tract. The name "Reo-" is derived from *respiratory enteric orphan viruses*. The term "orphan virus" refers to the fact that some of these viruses have been observed not associated with any known disease. Even though viruses in the Reoviridae family have more recently been identified with various diseases, the original name is still used.

Reovirus infection occurs often in humans, but most cases are mild or subclinical. Rotavirus, however, can cause severe diarrhea and intestinal distress in children. The virus can be readily detected in feces, and may also be recovered from pharyngeal or nasal secretions, urine, cerebrospinal fluid, and blood. Despite the ease of finding Reovirus in clinical specimens, their role in human disease or treatment is still uncertain.

Some viruses of this family, such as Phytoreovirus and Oryzavirus, infect plants. Most of the plant-infecting reoviruses are transmitted between plants by insect vectors. The viruses replicate in both the plant and the insect, generally causing disease in the plant, but little or no harm to the infected insect.

Recently researchers at the University of Leeds have found that Reovirus stimulates the body's own immune system to kill off the cancerous cells.

Structure

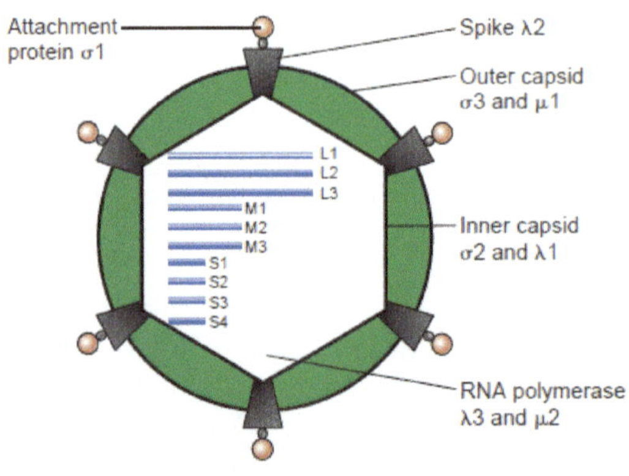

Structure of the reovirus virion

Reoviruses are non-enveloped and have an icosahedral capsid composed of an outer (T=13) and inner (T=2) protein shell. The genomes of viruses in Reoviridae contain 10–12 segments which are grouped into three categories corresponding to their size: L (large), M (medium) and S (small). Segments range from about 0.2 to 3 kbp and each segment encodes 1–3 proteins (10-14 proteins in total). Reoviridae proteins are denoted by the Greek character corresponding to the segment it was translated from (the L segment encodes for λ proteins, the M segment encodes for μ proteins and the S segment encodes for σ proteins).

Genus	Structure	Symmetry	Capsid	Genomic Arrangement	Genomic Segmentation
Aquareovirus	Icosahedral	T=13, T=2	Non-Enveloped	Linear	Segmented
Rotavirus	Icosahedral	T=13, T=2	Non-Enveloped	Linear	Segmented
Seadornavirus	Icosahedral	T=13, T=2	Non-Enveloped	Linear	Segmented
Coltivirus	Icosahedral	T=13, T=2	Non-Enveloped	Linear	Segmented
Fijivirus	Icosahedral	T=13, T=2	Non-Enveloped	Linear	Segmented
Phytoreovirus	Icosahedral	T=13, T=2	Non-Enveloped	Linear	Segmented
Mimoreovirus	Icosahedral	T=13, T=2	Non-Enveloped	Linear	Segmented
Orbivirus	Icosahedral	T=13, T=2	Non-Enveloped	Linear	Segmented
Cypovirus	Icosahedral	T=2	Non-Enveloped	Linear	Segmented
Orthoreovirus	Icosahedral	T=13, T=2	Non-Enveloped	Linear	Segmented
Idnoreovirus	Icosahedral	T=13, T=2	Non-Enveloped	Linear	Segmented
Dinovernavirus	Icosahedral	T=2	Non-Enveloped	Linear	Segmented
Oryzavirus	Icosahedral	T=13, T=2	Non-Enveloped	Linear	Segmented
Cardoreovirus	Icosahedral	T=13, T=2	Non-Enveloped	Linear	Segmented
Mycoreovirus	Icosahedral	T=13, T=2	Non-Enveloped	Linear	Segmented

Life Cycle

Viruses in the family Reoviridae have genomes consisting of segmented, double-stranded RNA (dsRNA). Because of this, replication occurs exclusively in the cytoplasm and the virus encodes several proteins which are needed for replication and conversion of the dsRNA genome into (+)-RNAs. The virus can enter the host cell via a receptor on the cell surface. The receptor is not known but is thought to include sialic acid and junctional adhesion molecules (JAMs). The virus is partially uncoated by proteases in the endolysosome, where the capsid is partially digested to allow further cell entry. The core particle then enters the cytoplasm by a yet unknown process where the genome is transcribed conservatively causing an excess of (+) sense strands, which are used as mRNA templates to synthesize (–) sense strands. Viral particles begin to assemble in the cytoplasm 6–7 hours after infection. Translation takes place by leaky scanning, suppression of termination, and ribosomal skipping. The virus exits the host cell by monopartite non-tubule guided viral movement, cell to cell movement, and existing in occlusion bodies after cell death and remaining infectious until finding another host.

Genus	Host Details	Tissue Tropism	Entry Details	Release Details	Replication Site	Assembly Site	Transmission
Aquareovirus	Aquatic vertebrates: fish; aquatic invertebrates: shellfish; aquatic invertebrates: crustaceans	None	Cell receptor endocytosis	Cell death	Cytoplasm	Cytoplasm	Passive diffusion

Rotavirus	Humans; vertebrates	Intestinal mucosa	Clathrin-mediated endocytosis	Cell death	Cytoplasm	Cytoplasm	Oral-fecal
Seadornavirus	Humans; cattle; pigs; mosquitoes	None	Cell receptor endocytosis	Cell death	Cytoplasm	Cytoplasm	Zoonosis; arthropod bite
Coltivirus	Humans; rodents; ticks; mosquitoes	Erythrocytes	Cell receptor endocytosis	Cell death	Cytoplasm	Cytoplasm	Arthropod bite
Fijivirus	Plants: gramineae; plants: liliacea; planthoppers	Phloem	Viral movement; mechanical inoculation	Cell death	Cytoplasm	Cytoplasm	Delphacid plant hoppers
Phytoreovirus	Oryza sativa; leafhoppers	Phloem	Viral movement; mechanical inoculation	Cell death	Cytoplasm	Cytoplasm	Leafhoppers
Mimoreovirus	Algae	None	Cell receptor endocytosis	Cell death	Cytoplasm	Cytoplasm	Arthropod bite
Orbivirus	Vertebrates; mosquitoes; midges; gnats; sandflies; ticks	None	Cell receptor endocytosis	Cell death	Cytoplasm	Cytoplasm	Arthropod bite
Cypovirus	Insects	Midgut; goblet; fat	Cell receptor endocytosis	Cell death	Cytoplasm	Cytoplasm	Polyhedra: oral-fecal; vertical: eggs
Orthoreovirus	Vertebrates	Epithelium: intestinal; epithelium:-bile duct; epithelium: lung; leukocytes; endothelium: CNS	Clathrin-mediated endocytosis	Cell death	Cytoplasm	Cytoplasm	Aerosol; oral-fecal
Idnoreovirus	Hymenoptera	Gut	Cell receptor endocytosis	Cell death	Cytoplasm	Cytoplasm	Unknown
Dinovernavirus	Insects; Mosquitoes	None	Unknown	Cell death	Cytoplasm	Cytoplasm	Unknown
Oryzavirus	Plants: graminae, Oryza sativa; planthoppers	None	Viral movement; mechanical inoculation	Cell death	Cytoplasm	Cytoplasm	Delphacid planthoppers
Cardoreovirus	Crustaceans: crabs	None	Cell receptor endocytosis	Cell death	Cytoplasm	Cytoplasm	Arthropod bite
Mycoreovirus	Fungi	Mycelium	Cell death; cytoplasmic exchange, sporogenesis; hyphal anastomosis	Cell death; cytoplasmic exchange, sporogenesis; hyphal anastomosis	Cytoplasm	Cytoplasm	Cytoplasmic exchange, sporogenesis; hyphal anastomosis

Multiplicity Reactivation

Multiplicity reactivation (MR) is the process by which 2 or more virus genomes, each containing inactivating genome damage, can interact within an infected cell to form a viable virus genome. McClain and Spendlove demonstrated MR for three types of reovirus after exposure to ultraviolet irradiation. In their experiments, reovirus particles were exposed to doses of UV-light that would be lethal in single infections. However, when two or more inactivated viruses were allowed to infect individual host cells MR occurred and viable progeny were produced. As they stated, multiplicity reactivation by definition involves some type of repair. Michod et al. reviewed numerous examples of MR in different viruses, and suggested that MR is a common form of sexual interaction in viruses that provides the benefit of recombinational repair of genome damages.

Taxonomy

The Reoviridae are divided into two subfamilies based on the presence of a "turret" protein on the inner capsid. From ICTV communications: "The name Spinareovirinae will be used to identify the subfamily containing the spiked or turreted viruses and is derived from 'reovirus' and the Latin word 'spina' as a prefix, which means spike, denoting the presence of spikes or turrets on the surface of the core particles. The term 'spiked' is an alternative to 'turreted', that was used in early research to describe the structure of the particle, particularly with the cypoviruses. The name Sedoreovirinae will be used to identify the subfamily containing the non-turreted virus genera and is derived from 'reovirus' and the Latin word 'sedo', which means smooth, denoting the absence of spikes or turrets from the core particles of these viruses, which have a relatively smooth morphology."

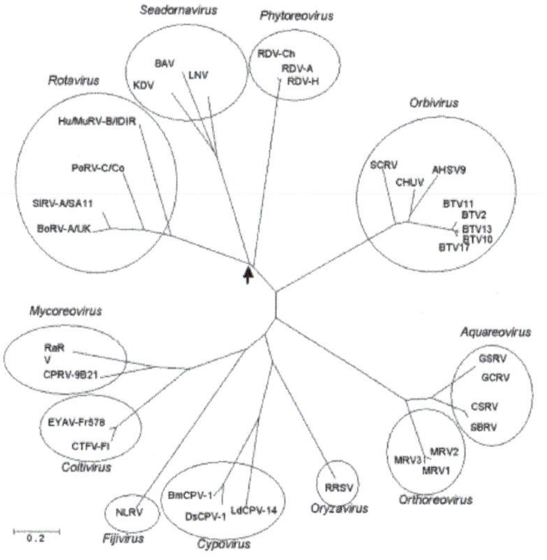

Phylogenetic comparison of the viral polymerase protein sequences of viruses of the family Reoviridae. Arrow indicates a common phylogenetic origin between seadornaviruses and rotaviruses.

Other Reoviridae

Additionally, there is one proposed genus that has not yet been accepted by the ICTV:

- Crabreovirus

Piscine reovirus is a Spinareovirinae-like virus that has not yet been classified.

Therapeutic Applications

Although reoviruses are mostly nonpathogenic in humans, these viruses have served as very productive experimental models for studies of viral pathogenesis. Newborn mice are exquisitely sensitive to reovirus infection and have been used as the preferred experimental system for studies of reovirus pathogenesis.

The reoviruses have been demonstrated to have oncolytic (cancer-killing) properties, encouraging the development of reovirus-based therapies for cancer treatment.

Reolysin is a formulation of reovirus that is currently in clinical trials for the treatment of various cancers.

Diseases Caused by Reoviruses

The family Reoviridae contains viruses which are most complex in nature. The term Reo stands for "respiratory enteric orphan," which was named because the first member was identified in the respiratory and the enteric tract of animals and humans and was not associated with any type of disease. They are generally spherical in shape and have icosahedral symmetry. They do not contain any envelope. The different viruses belonging to this family have their names which are indicative of their unique morphological features. For example in case of rotaviruses, rota stands for wheel like capsid with spikes, similarly in orbiviruses, orbi stands for ring shaped capsid. Some of the reoviruses are transmitted through the bite of female culicoides (blue tongue virus) or through tick bite (Colorado Tick fever).

The family *Reoviridae* contains six genera based on group specific antigen present on VP6 capsid protein

- Orthoreovirus
- Orbivirus
- Coltivirus
- Rotavirus
- Seadornavirus
- Aquareovirus

Table Diseases caused by Reoviruses

VIRUS	DISEASE
Orthoreovirus	Respiratory and enteric diseases
Orbivirus	Blue tongue fever in cattle and Sheep
	African Horse Sickness disease
Coltivirus	Colorado Tick fever
Rotavirus	Gastroenteritis and Diarrhoea
Aquareovirus	Diseases of Fish

Virion Property

Reovirus particles are non-enveloped, spherical having a diameter of approximately 85 nm. Genome contains a linear dsRNA divided into 10-12 segments. The overall genome size varies from 16 to 29 Kbp. They contain a cap at their 5' end while poly A tails are absent from 3' end.

Virus Replication

Virus replication takes place in the cytoplasm of the cell. Because of the segmented RNA genome chances of reassortment of genomic segments between different strains is very high. This results in genetic drift and shift leading to diversity among viruses which is reflected by numerous serotypes within each genus.

Virus enters the cell by receptor mediated endocytosis. The coated vesicle uncoats and fuses to the lysosomes under low pH condition. Virions are disrupted and viral inner core is released into the cytoplasm . Virus associated RNA polymerase utilizes the negative strand of each dsRNA segment as template to transcribe viral mRNA. These viral mRNA's are translated to form viral structural proteins that eventually assemble to form the infectious virion. Progeny viruses remain cell associated but are often released by lysis of the infected cell. Genomic RNA replication takes place within sub-viral particles in the cytoplasm of infected cells.

Important Reoviruses

Rotavirus

Rotaviruses are the most important human pathogens which lead to life threatening diarrhoea in young ones. It was first isolated from a children hospital in Australia. Rotavirus infection leads to destruction of intestinal villi. Transmission of the virus mainly takes place by fecal-oral route. Destruction of enterocyte (intestinal cells) causes mal-digestion and poor absorption of food. Intestinal epithelium of young ones has a slow turnover rate and that is the reason that rotavirus infection is more severe in case of infants. Rotavirus infection is characterized by severe diarrhoea, dehydration, weight loss and fatigue. Virus can be isolated in faeces in high amount. Enzyme immunoassay is commonly used to detect viral infection in infected individuals. Genetic reassortment vaccines are available for rotavirus infection.

African Horse Sickness

African horse sickness is caused by a member of genus orbivirus . It is a pantropic and fatal disease of horses, predominantly infecting endothelial cells and myocardium. Acute form of the disease is characterized by pneumonia, interlobular pulmonary oedema, pericarditis, haemorrhages and oedema of the visceral organs. The death can occur within 5 days in highly acute form of the disease. A more prolonged form of the sickness involves the cardiac vascular system. Mortality can be more than 80% depending on the immune status of the animal and virulence of the isolate. Subclinical disease can occur in donkeys and vaccinated horses. Diagnosis is usually carried out by using complement fixation tests and haemagglutination inhibition tests.

Bluetongue Virus

It is another important disease of livestock caused by a member of genus orbivirus . It causes high mortality in sheep and decrease in productivity of other farm animals. The virus is transmitted to the animals by the bite of Culicoides mosquitoes. The chances of outbreaks are more during the breeding season of Culicoides. Major signs of the disease include high fever, swelling of the face, hyperemia around the coronary band and cyanosis of the tongue (blue colour of the tongue). The disease can be restricted by controlling mosquitoes while polyvalent live vaccines are available for the animals.

Colorado Tick Fever

Colorado Tick fever is a disease caused by bite of tick (*Dermacentor andersoni*) infected with Coltivirus. Transmission of this disease is reported to be through blood transfusion from an infected individual. The virus usually infects and replicates in the erythrocyte. The symptoms of the disease include fever, headache, vomiting, abdominal pain, and encephalitis.

Retroviruses

Retroviruses are single stranded RNA viruses that undergo DNA phase during their replication. All the retroviruses contain reverse transcriptase enzyme and are capable of integrating with the host genome. Currently retroviruses are divided into 7 classes based on their biological and molecular properties. In general they are classified into three categories –

I. Oncoviruses

II. Lentiviruses

III. Spumaviruses

I. Oncoviruses

These viruses have long incubation period and are associated with immune deficiencies. Occasionally encephalopathy and arthritis may also be evident eg human immuno deficiency virus (HIV), feline immuno deficiency virus (FIV), and simian immuno deficiency virus (SIV).

II. Lentiviruses

These viruses have long incubation period and are associated with immune deficiencies. Ocasionally encephalopathy and arthritis may also be evident eg human immuno deficiency virus (HIV), feline immuno deficiency virus (FIV), simian immuno deficiency virus (SIV).

III. Spumaviruses

Spuma stands for foamy. The virus causes vacuolation and froth like appearance in infected cells. eg simian foamy virus (SFV).

General Concept

A. Out of three subfamilies of Retroviruses two infect humans namely oncornaviruses (Human T Cell Lymphotrophic Viruses (HTLV) I, II, and V) and lentiviruses (HIV 1, HIV 2).

HTLV-I causes cutaneous T-cell lymphomas,

HTLV-II causes hairy T-cell leukemias,

HTLV-V causes T-cell lymphomas and leukemias.

HIV-1 causes Acquired Immunodeficiency Syndrome (AIDS),

HIV-2 causes AIDS related syndrome in Africa.

B. Retroviruses contain RNA genome and an enzyme called reverse transcriptase, which makes circular DNA by using RNA as a template. The viral DNA integrates with the host cell chromosomes.

C. HIV is an enveloped virus which contains glycoprotein 120 (gp120) which binds to CD4 receptor present over the T helper cell.

D. Genome of HIV composed of two positive strands RNA which are capped at 5'end and polyadenylated at 3'end.

E. The tRNA acts as a primer for the synthesis of DNA using RNA as a template.

F. Long terminal repeats (LTR) present towards both the ends help in the integration of viral genome to host genome. It also serves as promoter and enhancer for the viral genome.

G. Group specific antigen (gag) helps in the formation of capsid protein. Polymerase (pol) encodes message for the formation of reverse transcriptase. Envelope (env) gene is associated with the formation of gp120 and gp41. tax/rex are the regions which encodes the factors involved in transactivation and other regulatory functions.

Morphology

Retroviruses are spherical and approximately 100 nm in size. The virion contains cone shaped nucleocapsid that encapsidates two copies of single stranded RNA about 10 Kb in size. In addition, capsid also contains three enzymes namely reverse transcriptase, proteases, and integrase. The nucleocapsid is covered by matrix protein which in turn is covered by an envelope containing two transmembrane glycoprotein called gp120 and gp 41.

Schematic representation of human immuno deficiency virus (HIV)

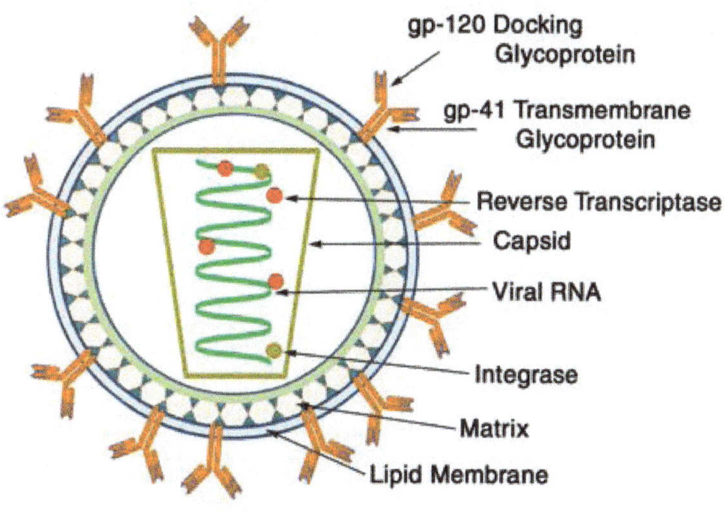

Virus Entry

HIV enters the host cell by infecting the T- helper cells and macrophages which contain CD4 receptor. The viral gp120 binds to CD4 receptor to initiate the entry process. Coreceptors such as chemokine receptors (CCR5 and CXCR4) also assist in entry process. Different HIV serotypes use different coreceptors. Following the binding, the virus and cell membrane fuses releasing the nucleocapsid inside the cell.

Virus Replication

Many of our present days understanding about the retrovirus comes from the work of HIV. Upon entry to the cells, viral RNA produces a polyprotein which is cleaved by virus encoded proteases to form individual protein. Virus uses the tRNA as a primer to synthesize DNA using reverse transcriptase enzyme. RNA from the hetroduplex form of RNA and DNA is then cleaved by RNase H. The remaining part of DNA is synthesized as double stranded and integrates into the host cell genome with the help of enzyme integrase. The viral genome remains integrated into the host genome and keeps on transcribing the viral RNA for the progeny virions. The progeny virions are assembled in the cytoplasm and exit by the process of budding taking the covering of the host cell membrane.

Schematic representation of HIV replication:

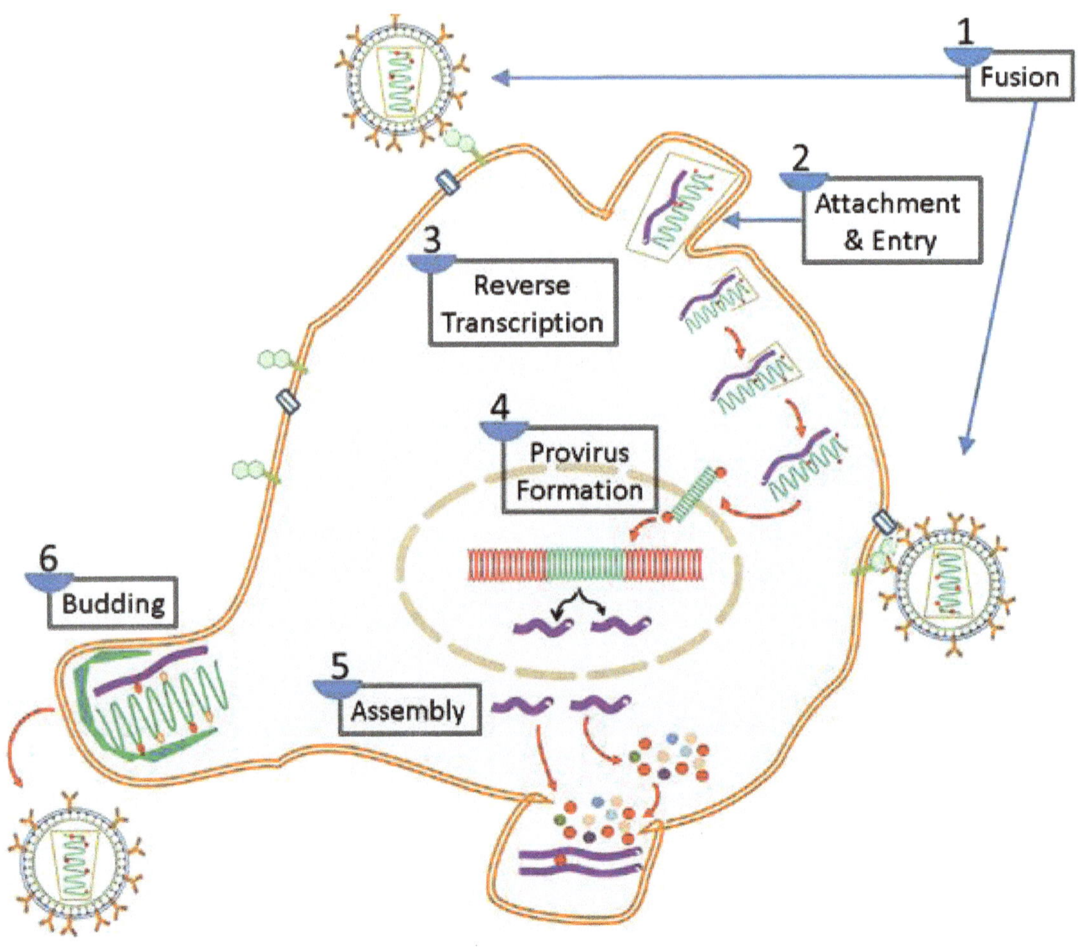

Endogenous Retroviruses

Genome of many vertebrate animals contains retroviral sequences. These sequences are mostly inactive or defective. On an average a normal human cell contains approximately 1, 00, 000 retroviral sequences. It is likely that the retroviruses can infect the germ lines early in the life, giving rise to these sequences in the human genome.

References

- Mettenleiter, T.C.; Sobrino, F., eds. (2008). Animal Viruses: Molecular Biology. Caister Academic. ISBN 978-1-904455-22-6
- Ghabrial SA (1998). "Origin, adaptation and evolutionary pathways of fungal viruses". Virus Genes. 16 (1): 119–31. doi:10.1023/A:1007966229595. PMID 9562896
- Patton JT, ed. (2008). Segmented Double-stranded RNA Viruses: Structure and Molecular Biology. Caister Academic Press. ISBN 978-1-904455-21-9
- Bouziat, R; et al. (April 7, 2017). "Reovirus infection triggers inflammatory responses to dietary antigens and development of celiac disease.". Science. 356 (6333): 44–50. doi:10.1126/science.aah5298
- Carter, John, Saunders ,Venetis (2007). Virology Principles and Applications. West Sussex PO19 8SQ, England: Jonn WILEY. p. 148. ISBN 978-0-470-02386-0
- Michod, R. E.; Bernstein, H.; Nedelcu, A. M. (2008). "Adaptive value of sex in microbial pathogens". Infection, Genetics and Evolution. 8 (3): 267–285. doi:10.1016/j.meegid.2008.01.002. PMID 18295550
- Hill C, Booth T, et al. (1999). "The structure of a cypovirus and the functional organization of dsRNA viruses". Nature Structural Biology. 6 (6): 565–9. doi:10.1038/9347. PMID 10360362

6

A Comprehensive Study of DNA Viruses

Since cell DNA takes place during the S phase only, the viruses that depend on the DNA polymerase of the host cell needs to wait to enter the S phase to regulate cell cycle. Large DNA viruses possess their own intermediate size DNA and DNA machinery. The aspects elucidated in this section are of vital importance, and provide a better understanding of DNA viruses.

Parvo- and Polyomaviruses

Parvo in latin stands for small. The virion is icosahedral, non-enveloped, and around 25 nm in diameter. These are the smallest of all animal viruses. They do not contain any viral or host enzyme and virion is made up of 80% protein and 20% DNA by weight.

The genome of parvovirus is linear, ssDNA which contains approximately 4500 to 5500 nucleotides. All parvoviruses contain terminal palindromic sequences at their 5' and 3' ends which allow the formation of Y or T shaped structures. In addition, they also contain inverted repeats at 3' and 5' ends which allow the circularization of the genome. Rep gene is required for replication of DNA while cap gene forms the capsid. The virion also contains 3 coat proteins namely, VP1, VP2, and VP3. VP3 is the most abundant among all three and it is made by the proteolytic cleavage of VP2. VP2 constitutes about 80% of capsid protein .

Genome map of parvoviruses:

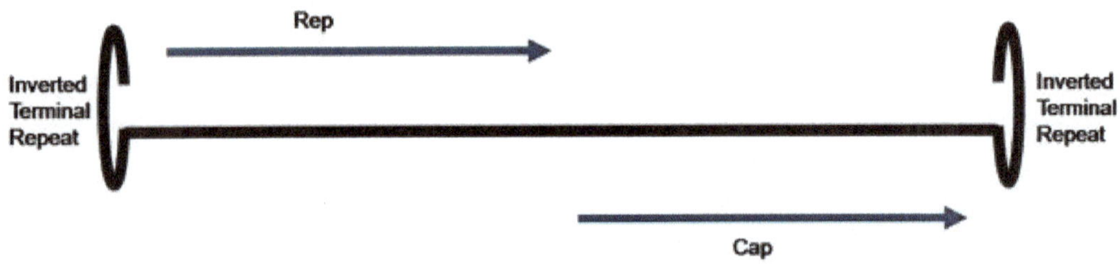

There are 3 genomes of Parvoviruses.-

1. Parvovirus
2. Dependovirus
3. Densovirus

1. Parvovirus – They only replicate in actively dividing cells. These viruses are highly resistant to heat, nucleases, detergents, proteases, and mild acid. They generally spread through body secretions. The most common strain of parvovirus that infects humans is B19.

2. Dependovirus – They are also called as adeno associated virus (AAV). They require adenoviruses to replicate. Upon infection in absence of helper virus they can establish latent infection by integrating into the host genome. They are one of the very important tools for targeted gene therapy viral vector.

Helper dependent replication strategy of adeno-associated viruses:

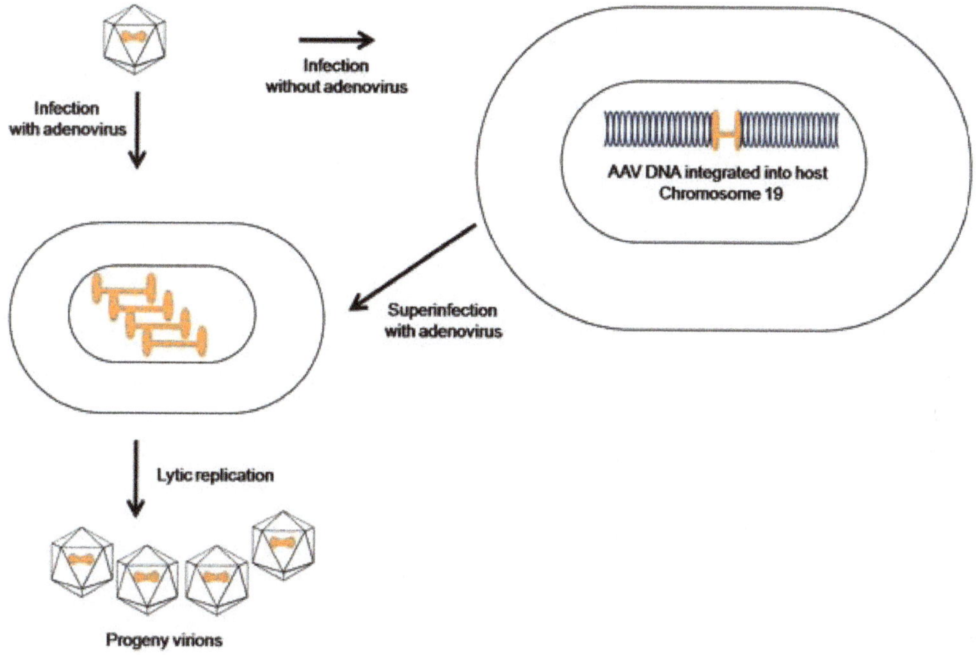

3. Densovirus – They infect only invertebrates.

Replication

The receptor for parvovirus has not been identified yet. It replicates with the help of host DNA polymerase and its assembly occurs in nucleus. Palindromic sequences at the termini are the initiation sites for DNA replication. Following replication, transcription is carried out by host cell RNA polymerase II. The genome of parvovirus codes for two non- structural proteins NS1 and NS2 apart from 3 coat proteins VP1, VP2 and VP3. Translation of the mRNA occurs in the cytoplasm and protein enters into the nucleus where assembly occurs. The virion exits the cell by lysis and the whole process is completed in 24hrs.

Papovaviruses

The name is derived from *pa*pillomas (warts), po lyomas (multifocal tumors), and vacuoles in infected cells. These viruses contain dsDNA and are non- enveloped and spherical. They replicate and assemble in the nucleus of the infected cell and are released out following the lysis of the cell.

There are two genera

1) Papilloma

2) Polyoma

Papilloma - Genome is around 8000 bp long. They depend upon the replication machinery of the host cell. They often infect basal cell layers of the skin and hence are associated with warts. Papilloma viruses do not grow in tissue culture.

Polyoma - Genome is around 6000 bp long. They are associated with leukoencephalopathy and immunosuppression in humans. The most common virus SV40 that has been used to study mammalian replication belongs to this group. SV 40 makes large and small T-antigens which are required for viral DNA replication as it binds to origin of replication and is known to possess helicase and ATPase activity.

Replication

- Adsorption of virions to the cell surface and entry by endocytosis.
- Transport to the cell nucleus and uncoating
- Transcription to produce early gene mRNAs and translation to produce early proteins (T antigens).
- Viral DNA replication and transcription of late gene mRNAs.
- Translation to produce late proteins (capsid proteins) and assembly of progeny virions in the nucleus.
- Release of virions from the cell

General replication strategies of polyoma and papillomaviruses:

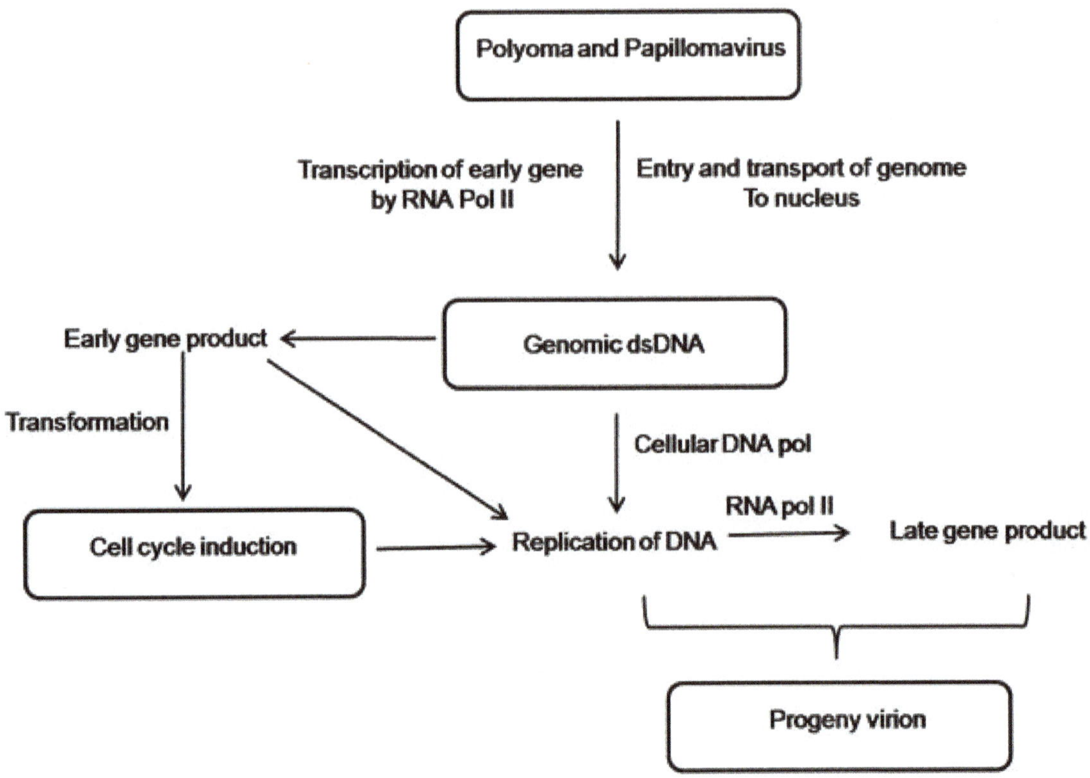

Large DNA Viruses

The viruses with large DNA have different strategies for their genome replication. The virus having intermediate size DNA (Adenoviruses) carry their own DNA replication machinery including DNA polymerase and other regulatory proteins. However, they still depend on the host cell RNA pol-II for the transcription of the viral RNA. The virus having large genome size like pox and herpesviruses have their own machinery to fulfill the requirement of RNA transcription and genomic DNA replication.

Virus type	Genome	Replication	Example
Adenovirus	dsDNA, 35 kb	Nucleus with DNA polymerase	Many serotypes infecting humans as well as animals
Herpesvirus	dsDNA, 120-230 kbp	Nucleus with DNA polymerase	HSV-1 and -2, Varicella/Zoster, HHV-5, 6, &7, Epstein-Barr
Poxvirus	dsDNA, 200 kbp	cytoplasm with DNA polymerase	Smallpox, Sheeppox, ORF, cowpox, vaccinia

General replication strategies of Adenoviruses

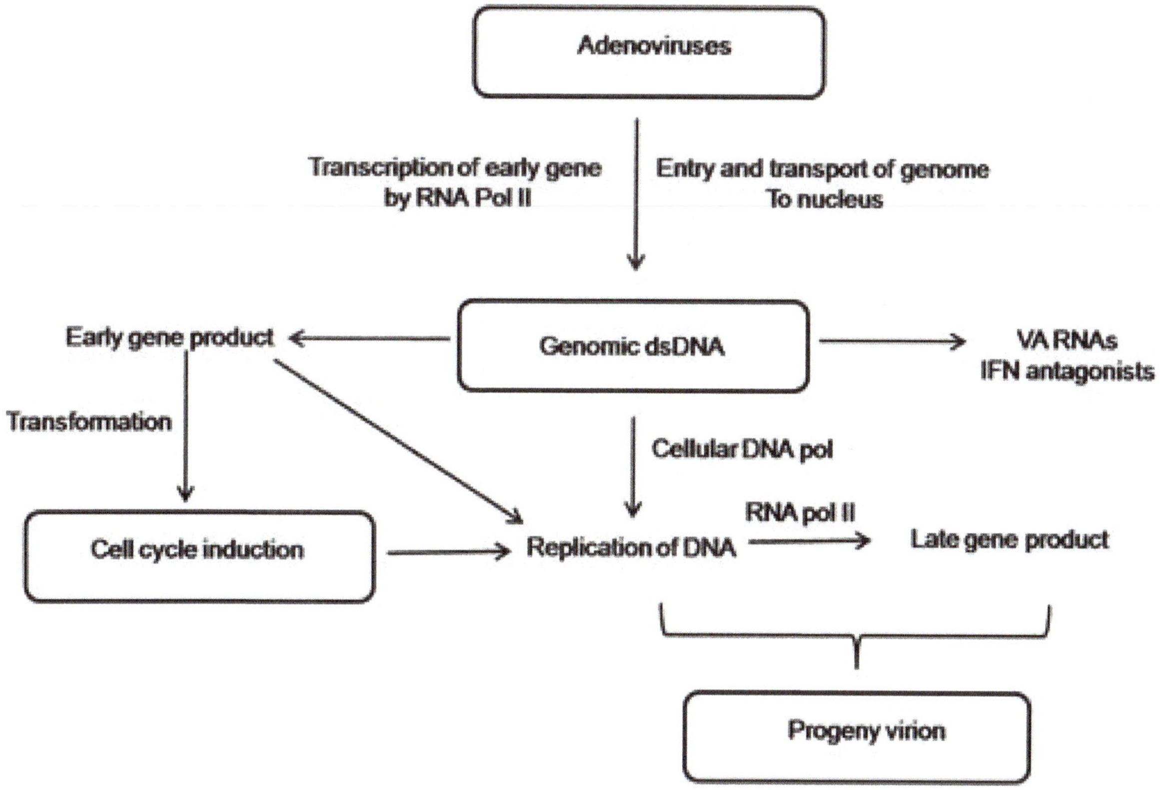

General replication strategies of herpesviruses

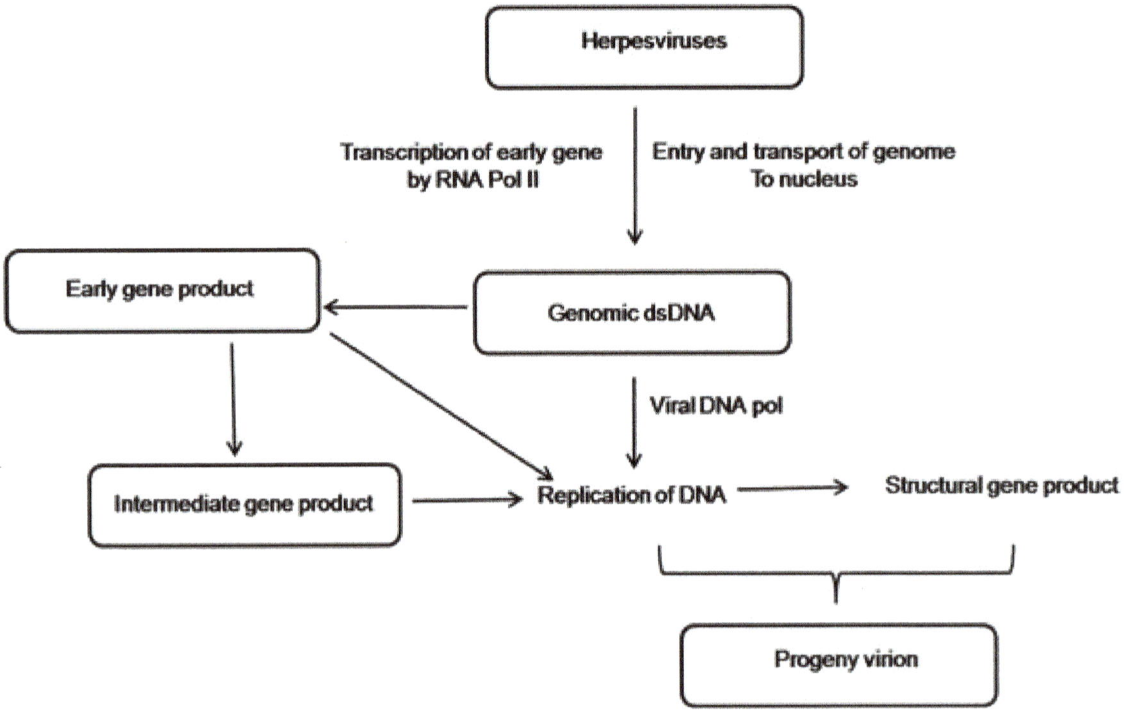

Oncogenic DNA Viruses

Epstein-Barr virus, *Hepatitis B* virus, papillomaviruses and Human Herpesvirus type 8 (HHV-8) or Kaposi's sarcoma are associated with one or other type of cancer in humans. In addition, HIV induces severe immunosuppression and facilitates development of cancer by other persisting infections, especially by HHV-8, Epstein-Barr virus and human papillomaviruses. Thus these agents contribute indirectly to human cancer. The mechanism about how these DNA viruses induce cancer is quite well known now. E6 and E7 gene of human papillomavirus modulate large array of cellular gene making cells vulnerable to undergo uncontrolled multiplication. Similarly, Epstein-Barr virus nuclear antigen 2 (EBNA-2) results in the induction of viral oncogenes that modulate many cell proteins. Many liver associated malignancies are caused by Hepatitis B virus. Bovine papillomaviruses can induce tumors (sarcoids) in horses and donkeys.

Herpesviridae

Herpesviridae is a large family of DNA viruses that cause diseases in animals, including humans. The members of this family are also known as herpesviruses. The family name is derived from the Greek word *herpein* ("to creep"), referring to the latent, recurring infections typical of this group of viruses. *Herpesviridae* can cause latent or lytic infections.

At least five species of Herpesviridae – HSV-1 and HSV-2 (both of which can cause orolabial herpes and genital herpes), varicella zoster virus (the cause of chickenpox and shingles), Epstein-Barr virus (implicated in several diseases, including mononucleosis and some cancers), and cytomegalovirus – are extremely widespread among humans. More than 90% of adults have been infected with at least one of these, and a latent form of the virus remains in most people.

There are 9 herpesvirus types known to infect humans: herpes simplex viruses 1 and 2, HSV-1 and HSV-2, (also known as HHV1 and HHV2), varicella-zoster virus (VZV, which may also be called by its ICTV name, HHV-3), Epstein-Barr virus (EBV or HHV-4), human cytomegalovirus (HCMV or HHV-5), human herpesvirus 6A and 6B (HHV-6A and HHV-6B), human herpesvirus 7 (HHV-7), and Kaposi's sarcoma-associated herpesvirus (KSHV, also known as HHV-8). In total, there are more than 130 herpesviruses, some of them from mammals, birds, fish, reptiles, amphibians, and mollusks.

Viral Structure

Herpesviruses all share a common structure—all herpesviruses are composed of relatively large double-stranded, linear DNA genomes encoding 100-200 genes encased within an icosahedral protein cage called the capsid which is itself wrapped in a protein layer called the tegument containing both viral proteins and viral mRNAs and a lipid bilayer membrane called the envelope. This whole particle is known as a virion.

Genus	Structure	Symmetry	Capsid	Genomic Arrangement	Genomic Segmentation
Iltovirus	Spherical Pleomorphic	T=16	Enveloped	Linear	Monopartite
Proboscivirus	Spherical Pleomorphic	T=16	Enveloped	Linear	Monopartite
Cytomegalovirus	Spherical Pleomorphic	T=16	Enveloped	Linear	Monopartite
Mardivirus	Spherical Pleomorphic	T=16	Enveloped	Linear	Monopartite
Rhadinovirus	Spherical Pleomorphic	T=16	Enveloped	Linear	Monopartite
Macavirus	Spherical Pleomorphic	T=16	Enveloped	Linear	Monopartite
Roseolovirus	Spherical Pleomorphic	T=16	Enveloped	Linear	Monopartite
Simplexvirus	Spherical Pleomorphic	T=16	Enveloped	Linear	Monopartite
Scutavirus	Spherical Pleomorphic	T=16	Enveloped	Linear	Monopartite
Varicellovirus	Spherical Pleomorphic	T=16	Enveloped	Linear	Monopartite
Percavirus	Spherical Pleomorphic	T=16	Enveloped	Linear	Monopartite
Lymphocryptovirus	Spherical Pleomorphic	T=16	Enveloped	Linear	Monopartite
Muromegalovirus	Spherical Pleomorphic	T=16	Enveloped	Linear	Monopartite

Herpesvirus Life Cycle

All herpesviruses are *nuclear-replicating*—the viral DNA is transcribed to mRNA within the infected cell's nucleus.

Infection is initiated when a viral particle contacts a cell with specific types of receptor molecules on the cell surface. Following binding of viral envelope glycoproteins to cell membrane receptors, the virion is internalized and dismantled, allowing viral DNA to migrate to the cell nucleus. Within the nucleus, replication of viral DNA and transcription of viral genes occurs.

During symptomatic infection, infected cells transcribe *lytic* viral genes. In some host cells, a small number of viral genes termed *latency associated transcript* (LAT) accumulate instead. In this fashion the virus can persist in the cell (and thus the host) indefinitely. While primary infection is often accompanied by a self-limited period of clinical illness, long-term latency is symptom-free.

Reactivation of latent viruses has been implicated in a number of diseases (e.g. Shingles, *Pityriasis Rosea*). Following activation, transcription of viral genes transitions from latency-associated LAT to multiple *lytic* genes; these lead to enhanced replication and virus production. Often, lytic activation leads to cell death. Clinically, lytic activation is often accompanied by emergence of non-specific symptoms such as low grade fever, headache, sore throat, malaise, and rash as well as clinical signs such as swollen or tender lymph nodes and immunological findings such as reduced levels of natural killer cells.

Genus	Host Details	Tissue Tropism	Entry Details	Release Details	Replication Site	Assembly Site	Transmission
Iltovirus	Birds: galliform: psittacine	None	Cell receptor endocytosis	Budding	Nucleus	Nucleus	Oral-fecal; aerosol
Proboscivirus	Elephants	None	Glycoproteins	Budding	Nucleus	Nucleus	Contact
Cytomegalovirus	Humans; monkeys	Epithelial mucosa	Glycoproteins	Budding	Nucleus	Nucleus	Urine; saliva
Mardivirus	Chickens; turkeys; quail	None	Cell receptor endocytosis	Budding	Nucleus	Nucleus	Aerosol
Rhadinovirus	Humans; mammals	B-lymphocytes	Glycoproteins	Budding	Nucleus	Nucleus	Sex; saliva
Macavirus	Mammals	B-lymphocytes	Glycoproteins	Budding	Nucleus	Nucleus	Sex; saliva
Roseolovirus	Humans	T-cells; B-cells; NK-cell; monocytes; macrophages; epithelial	Glycoproteins	Budding	Nucleus	Nucleus	Respiratory contact
Simplexvirus	Humans; mammals	Epithelial mucosa	Cell receptor endocytosis	Budding	Nucleus	Nucleus	Saliva
Scutavirus	Sea turtles	None	Cell receptor endocytosis	Budding	Nucleus	Nucleus	Aerosol
Varicellovirus	Mammals	Epithelial mucosa	Glycoproteins	Budding	Nucleus	Nucleus	Aerosol
Percavirus	Mammals	B-lymphocytes	Glycoproteins	Budding	Nucleus	Nucleus	Sex; saliva
Lymphocryptovirus	Humans; mammals	B-lymphocytes	Glycoproteins	Budding	Nucleus	Nucleus	Saliva
Muromegalovirus	Rodents	Salivary glands	Glycoproteins	Budding	Nucleus	Nucleus	Contact

Taxonomy

The herpesvirus was first isolated from the blue wildebeest in 1960 by veterinary scientist Walter Plowright. The genus *Herpesvirus* was established in 1971 in the first report of the International Committee on Taxonomy of Viruses (ICTV). This genus consisted of 23 viruses and 4 groups of viruses. In the second ICTV report in 1976 this genus was elevated to family level - the *Herpetoviridae.* Because of possible confusion with viruses derived from reptiles this name was changed in the third report in 1979 to *Herpesviridae.* In this report the family *Herpesviridae* was divided into 3 subfamilies (*Alphaherpesvirinae*, *Betaherpesvirinae* and *Gammaherpesvirinae*) and 5 unnamed genera: 21 viruses were listed. In 2009 the family *Herpesviridae* was elevated to the order *Herpesvirales.* This elevation was necessitated by the discovery that the herpes viruses of fish and molluscs were only distantly related to those of birds and mammals. Two new families were created - the family *Alloherpesviridae* which incorporates bony fish and frog viruses and the family *Malacoherpesviridae* which contains those of molluscs.

This order currently has 3 families, 3 subfamilies plus 1 unassigned, 17 genera, 90 species and plus 48 as yet unassigned viruses.

Virus Naming System

The system of naming herpes viruses was originated in 1973 and has been elaborated considerably since. The recommended naming system specified that each herpes virus should be named after the taxon (family or subfamily) to which its primary natural host belongs. The subfamily name is used for viruses from members of the family *Bovidae* or from primates (the virus name ending in –ine, e.g. bovine) and the host family name for other viruses (ending in –id, e.g. equid). Human herpes viruses have been treated as an exception (*human* rather than *hominid*). Following the host-derived term, the word *herpes virus* is added, followed by an Arabic numeral (1,2,3,...). These last two additions bear no implied meaning about taxonomic or biological properties of the virus.

Some exceptions to this system exist. A number of viruses' names (e.g. Epstein–Barr virus) are so widely used that it is impractical to attempt to insist on their replacement. This has led to a dual nomenclature in the literature for some herpes viruses. All herpes viruses described since this system was adopted have been named in accordance with it.

Evolution

The three mammalian subfamilies - Alpha, Beta and Gamma - arose approximately 180 million years ago to 220 million years ago. The major sublineages within these subfamilies were probably generated before the mammalian radiation of 80 million years ago to 60 million years ago. Speciations within sublineages took place in the last 80 million years probably with a major component of cospeciation with host lineages.

All the currently known bird and reptile species are alphaherpesviruses. Although the branching order of the herpes viruses has not yet been resolved, because herpes viruses and their hosts tend to coevolve this is suggestive that the alphaherpesviruses may have been the earliest branch.

The date of evolution of the iltovirus genus has been estimated to be 200 million years ago while those of the mardivirus and simplex genera have been estimated to be between 150 million years ago and 100 million years ago.

Immune System Evasions

Herpesviruses are known for their ability to establish lifelong infections. One way this is possible is through immune evasion. Herpesviruses have many different ways of evading the immune system. One such way is by encoding a protein mimicking human interleukin 10 (hIL-10) and another is by downregulation of the major histocompatibility complex II (MHC II) in infected cells.

cmvIL-10

Research conducted on cytomegalovirus (CMV) indicates that the viral human IL-10 homolog, cmvIL-10, is important in inhibiting pro-inflammatory cytokine synthesis. The cmvIL-10 protein has 27% identity with hIL-10 and only one conserved residue out of the nine amino acids that make up the functional site for cytokine synthesis inhibition on hIL-10. There is, however, much similarity in the functions of hIL-10 and cmvIL-10. Both have been shown to down regulate IFN-γ, IL-1α, GM-CSF, IL-6 and TNF-α, which are all pro-inflammatory cytokines. They have also been shown to play a role in downregulating MHC I and MHC II and up regulating HLA-G (non-classical MHC I). These two events allow for immune evasion by suppressing the cell-mediated immune response and natural killer cell response, respectively. The similarities between hIL-10 and cmvIL-10 may be explained by the fact that hIL-10 and cmvIL-10 both use the same cell surface receptor, the hIL-10 receptor. One difference in the function of hIL-10 and cmvIL-10 is that hIL-10 causes human peripheral blood mononuclear cells (PBMC) to both increase and decrease in proliferation whereas cmvIL-10 only causes a decrease in proliferation of PBMCs. This indicates that cmvIL-10 may lack the stimulatory effects that hIL-10 has on these cells.

It was found that cmvIL-10 functions through phosphorylation of the Stat3 protein. It was originally thought that this phosphorylation was a result of the JAK-STAT pathway. However, despite evidence that JAK does indeed phosphorylate Stat3, its inhibition has no significant influence on cytokine synthesis inhibition. Another protein, PI3K, was also found to phosphorylate Stat3. PI3K inhibition, unlike JAK inhibition, did have a significant impact on cytokine synthesis. The difference between PI3K and JAK in Stat3 phosphorylation is that PI3K phosphorylates Stat3 on the S727 residue whereas JAK phosphorylates Stat3 on the Y705 residue. This difference in phosphorylation positions seems to be the key factor in Stat3 activation leading to inhibition of pro-inflammatory cytokine synthesis. In fact, when a PI3K inhibitor is added to cells, the cytokine synthesis levels are significantly restored. The fact that cytokine levels are not completely restored indicates there is another pathway activated by cmvIL-10 that is inhibiting cytokine system nthesis. The proposed mechanism is that cmvIL-10 activates PI3K which in turn activates PKB (Akt). PKB may then activate mTOR, which may target Stat3 for phosphorylation on the S727 residue.

MHC Downregulation

Another one of the many ways in which herpes viruses evade the immune system is by down regulation of MHC I and MHC II. This is observed in almost every human herpesvirus. Down regulation of MHC I and MHC II can come about by many different mechanisms, most causing the MHC to be absent from the cell surface. As discussed above, one way is by a viral chemokine homolog such as IL-10. Another mechanism to down regulate MHCs is to encode viral proteins that detain the newly formed MHC in the endoplasmic reticulum (ER). The MHC cannot reach the cell surface and therefore cannot activate the T cell response. The MHCs can also be targeted for destruction in the proteasome or lysosome. The ER protein TAP also plays a role in MHC down regulation. Viral proteins inhibit TAP preventing the MHC from picking up a viral antigen peptide. This prevents proper folding of the MHC and therefore the MHC does not reach the cell surface.

It is important to note that HLA-G is often up regulated in addition to downregulation of MHC I and MHC II. This prevents the natural killer cell response.

Human Herpesvirus Types

There are nine distinct viruses in this family known to cause disease in humans.

Human Herpesvirus (HHV) classification						
Name	**Synonym**	**Subfamily**	**Primary Target Cell**	**Pathophysiology**	**Site of Latency**	**Means of Spread**
HHV1	Herpes simplex virus-1 (HSV-1)	α (Alpha)	Mucoepithelial	Oral and/or genital herpes (predominantly orofacial), as well as other herpes simplex infections	Neuron	Close contact (oral or sexually transmitted infection)
HHV-2	Herpes simplex virus-2 (HSV-2)	α	Mucoepithelial	Oral and/or genital herpes (predominantly genital), as well as other herpes simplex infections	Neuron	Close contact (oral or sexually transmitted disease)
HHV-3	Varicella zoster virus (VZV)	α	Mucoepithelial	Chickenpox and shingles	Neuron	Respiratory and close contact (including sexually transmitted disease)
HHV-4	Epstein-Barr virus (EBV), lymphocryptovirus	γ (Gamma)	B cells and epithelial cells	Infectious mononucleosis, Burkitt's lymphoma, CNS lymphoma in AIDS patients, post-transplant lymphoproliferative syndrome (PTLD), nasopharyngeal carcinoma, HIV-associated hairy leukoplakia	B cell	Close contact, transfusions, tissue transplant, and congenital
HHV-5	Cytomegalovirus (CMV)	β (Beta)	Monocyte, lymphocyte, and epithelial cells	Infectious mononucleosis-like syndrome, retinitis	Monocyte, lymphocyte, and ?	Saliva, urine, blood, breast milk
HHV-6A and 6B	Roseolovirus, Herpes lymphotropic virus	β	T cells and ?	*Sixth disease* (roseola infantum or *exanthem subitum*)	T cells and ?	Respiratory and close contact?

HHV-7		β	T cells and ?	(pytiriasis rosea, roseola infantum or *exanthem subitum*)	T cells and ?	?
HHV-8	Kaposi's sarcoma-associated herpesvirus (KSHV), a type of rhadinovirus	γ	Lymphocyte and other cells	Kaposi's sarcoma, primary effusion lymphoma, some types of multicentric Castleman's disease	B cell	Close contact (sexual), saliva?

Zoonotic Herpesviruses

In addition to the herpesviruses considered endemic in humans, some viruses associated primarily with animals may infect humans. These are zoonotic infections:

Zoonotic Herpesviruses				
Species	**Type**	**Synonym**	**Sub-family**	**Human Pathophysiology**
Macaque monkey	CeHV-1	Cercopithecine herpesvirus-1, (monkey B virus)	α	Very unusual, with only approximately 25 human cases reported. Untreated infection is often deadly; sixteen of the 25 cases resulted in fatal encephalomyelitis. At least four cases resulted in survival with severe neurologic impairment. Symptom awareness and early treatment are important for laboratory workers facing exposure.
Mouse	MuHV4	Murid herpesvirus 68 (MHV-68)	γ	Zoonotic infection found in 4.5% of general population and more common in laboratory workers handling infected mice. ELISA tests show factor-of-four (x4) false positive results, due to antibody cross-reaction with other Herpes viruses.

Animal Herpesviruses

In animal virology, the most important herpesviruses belong to the subfamily *Alphaherpesvirinae*. Research on pseudorabies virus (PrV), the causative agent of Aujeszky's disease in pigs, has pioneered animal disease control with genetically modified vaccines. PrV is now extensively studied as a model for basic processes during lytic herpesvirus infection, and for unraveling molecular mechanisms of herpesvirus neurotropism, whereas bovine herpesvirus 1, the causative agent of bovine infectious rhinotracheitis and pustular vulvovaginitis, is analyzed to elucidate molecular mechanisms of latency. The avian infectious laryngotracheitis virus is phylogenetically distant from these two viruses and serves to underline similarity and diversity within the *Alphaherpesvirinae*.

Herpesvirus Virion

Herpesviruses have a complex structure because of its large size, multiple proteins, and tegument. The viral genome is a linear dsDNA of about 125-250 Kbp. The DNA is encapsidated inside an icosahedral capsid which is surrounded by a tegument. The tegument of herpesvirus contains many proteins while envelope is composed of 10 or more glycoproteins. The structural proteins of the herpesviruses are called as viral protein (VP) and VP5 is the most abundant protein present in the capsid. The envelope glycoproteins such as gB, gC and gD are the antigenic determinants and are involved in mounting the host immune response. In addition, the virion also contains the hexon and penton fibers. Genome of the herpesvirus contains two unique sequences; large and small and both are flanked by the repeat sequences. The genome encodes more than 75 proteins and many mRNA subspecies. Both strands of the DNA are used for the coding purpose. As some genes are present in the inverted repeats so the genome contains a pair of those genes in each strand.

Herpesvirus Replication

Herpesvirus largely infects humans, many animals and lab animal in research laboratories. The virus binds first to heparan sulfate and then to cell adhesion molecules such as nectins . The virus then fuses with the cell membrane and enters the cell following endocytosis . The nucleocapsid and the tegument are released into the cytoplasm. The replication of the genome occurs in the nucleus and therefore nucleocapsid is first transported into the nucleus. The linear DNA molecule is converted into closed and circular in the nucleus of the host cells. The closed circular DNA then binds with the histones. The major tegument protein VP16 helps in modulating the viral gene expression and is transported along with the viral DNA to the nucleus. The herpesvirus genes express as immediate early (IE), early (E), and late (L). VP16 acts as a transcription factor to recruit RNA polymerase II to activate the immediate early genes. Early proteins have their role in viral DNA replication and late proteins are formed during the assembly of the virion. Most of the late proteins are structural proteins. DNA replication in herpesvirus starts with a θ mode and later switches to rolling circle mode. Rolling circle mode of replication is the major form observed in the herpesviruses.

Rolling circle replication in herpesviruses

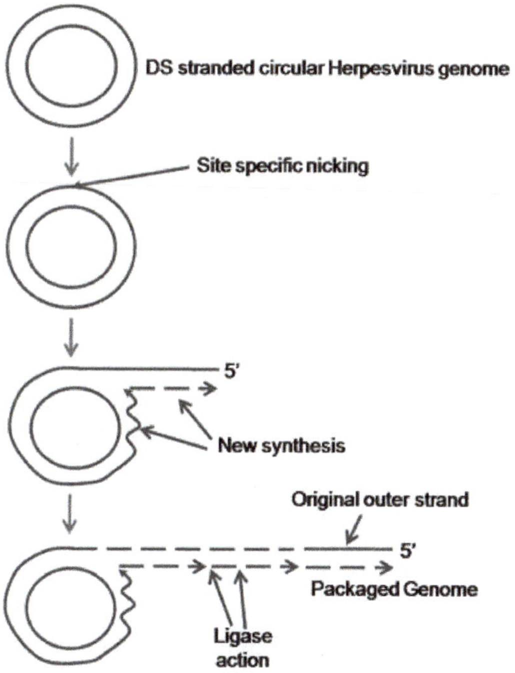

Latency During Herpesvirus Infections

- After entry virus releases many factors that initiate infection. The factors include virion host shut-off (VHS), α-trans inducing factor (αTIF) and many others.

α-TIF- helps in the synthesis of 5α mRNA.

VHS- favors to shut off host protein synthesis by degrading cellular mRNA.

- The viral genome and αTIF migrates to the nucleus where viral gene expression begins.
- New viruses bud out from the cell and infect other neighboring epithelial cells.
- Some newly formed virions cross the synapses and travel downwards the axon to the nerve cells towards peripheral ganglion.
- The virions become latent inside neurons with the expression of Latency associated transcript (LAT).

  LAT- It is an mRNA made during latency by viruses.
- During the process of reactivation new viruses are made in the nerve cell. They travel back downwards the axon to infect the epithelial cells again.
- Nerve cells in the ganglion are well connected to the brain but virus rarely goes in that direction.

Important Herpesviruses

Herpes Simplex Viruses 1 and 2

Herpes simplex viruses 1 and 2 (HSV-1 and HSV-2) infect epithelial cells of the buccal cavity, genital mucosa membrane, skin and cornea. Generally the virus migrates to central nervous system via neurons and initiates a latent infection. HSV-1 is mostly transmitted by lips and nasal contacts mostly to the young ones (1-2 years). HSV-2 is mostly transmitted by sexual contact and hence called as genital herpes.

Varicella-zoster Virus

Varicella-zoster virus leads to a condition commonly called as chicken pox (varicella) where virus spreads to the skin and produces rashes. The rashes are mostly towards the face and trunk area. It may spread to CNS to produce a latent infection and reactivates during stress or administration of corticosteroids leading to a condition called as shingles . The symptoms include rashes in different body parts, fever, headache, joint pain, and swollen lymph nodes.

Epstein-Barr Virus

Epstein-Barr virus is generally transmitted by saliva from an infected individual and spread in the body by its multiplication in the B cells. The virus infects the young ones with asymptomatic infection which activates during adolescence. The virus leads to a condition called as infectious mononucleosis or glandular fever. Epstein-Barr virus is also associated with different kind of cancers in humans. In medical science the infection is referred as kissing disease since it is transmitted by kissing through saliva.

Human Cytomegalovirus

Human cytomegalovirus is transmitted vertically from mother to foetus. The infection at birth can cause reduced brain size and enlargement of the liver and spleen. During the later phase of life virus can cause hearing loss and mental retardation. HIV positive patients who are immunocompromised can easily be infected by cytomegalovirus which terminates into life threatening pneumonia or hepatitis.

Adenoviruses

Adenoviruses are one of the major causative agents of upper respiratory tract or common cold infection in humans. In addition, they also cause conjunctivitis (eye inflammation), tonsillitis (inflammation of tonsils), gastroenteritis (inflammation of intestine), urinary tract infections, and infection to brain.

- There are four genera of adenoviruses
- Aviadenovirus- Infecting to avian species
- Mastadenovirus- Infecting to mammals
- Atadenovirus- Infecting to avian and humans
- Siadenovirus- Infecting to avian, mammals, and reptiles.

Adenovirus Structure and Genome Organization

Adenoviruses are non-enveloped and icosahedral particles. They are 60-90 nm in diameter and contain 252 capsomers (240 hexons and 12 pentons) in the vertices of the icosahedrons.

The hexons on the virions are involved in the stabilization and assembly of the viral particle.

They contain a penton fiber that projects from each apex from the virion surface. The penton fiber consists of a shaft and a globular head. They are involved in the attachment of the virus to the surface of the host cell. They are very fragile and usually detached during preparation for electron microscopy.

Schematic representation of adenovirus virion

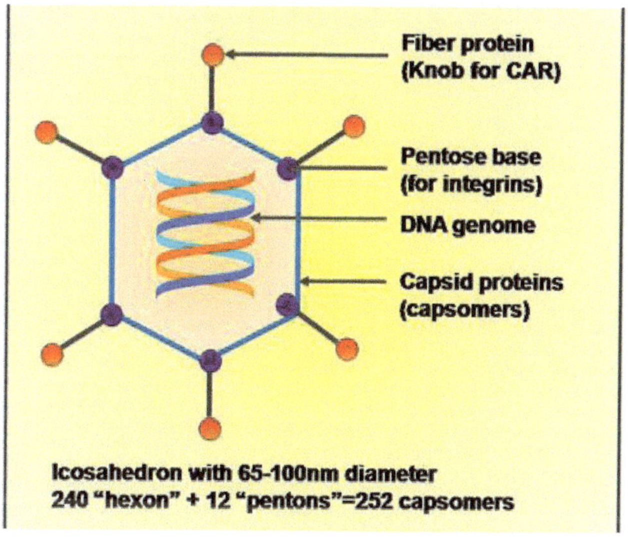

Genome of adenovirus contains linear dsDNA of about 35 kb which encodes approximately 40 different proteins. The genome has inverted terminal repeats which are required during the replication process. The adenovirus DNA contains terminal protein at its 5' end. The early genes (E1-E4) are present towards either of the ends and are required to control the transcription and viral DNA replication. The late genes are generally associated with the viral structural proteins.

Adenovirus genome

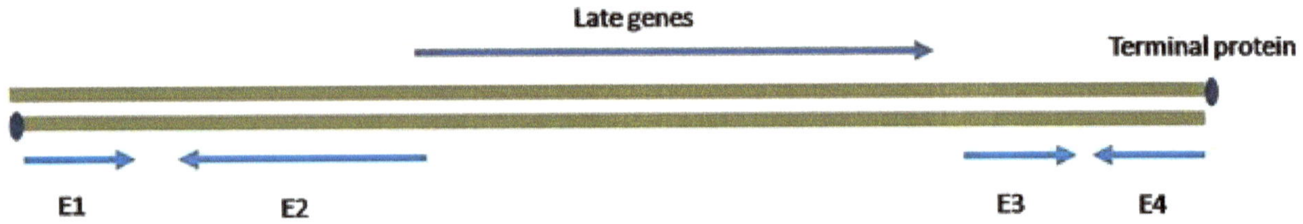

Adenovirus Replication

To enter the cells they use a receptor present in the host cell called as CAR (coxsackie and adenovirus receptor). The internalization of the virus particles occur through receptor mediated endocytosis. After entry the endosome containing the virus particle migrates to nucleus and the genetic information of the virus is released into the nucleus. Transcription of the first gene is done by the terminal protein attached with the viral DNA. Viral mRNA is then transported to the cytoplasm and translated into the viral proteins. Virus assembly takes place in the cytoplasm and the mature viral particles get released from the infected cells after killing them by accumulated adenoviral death proteins.

Adenovirus life cycle

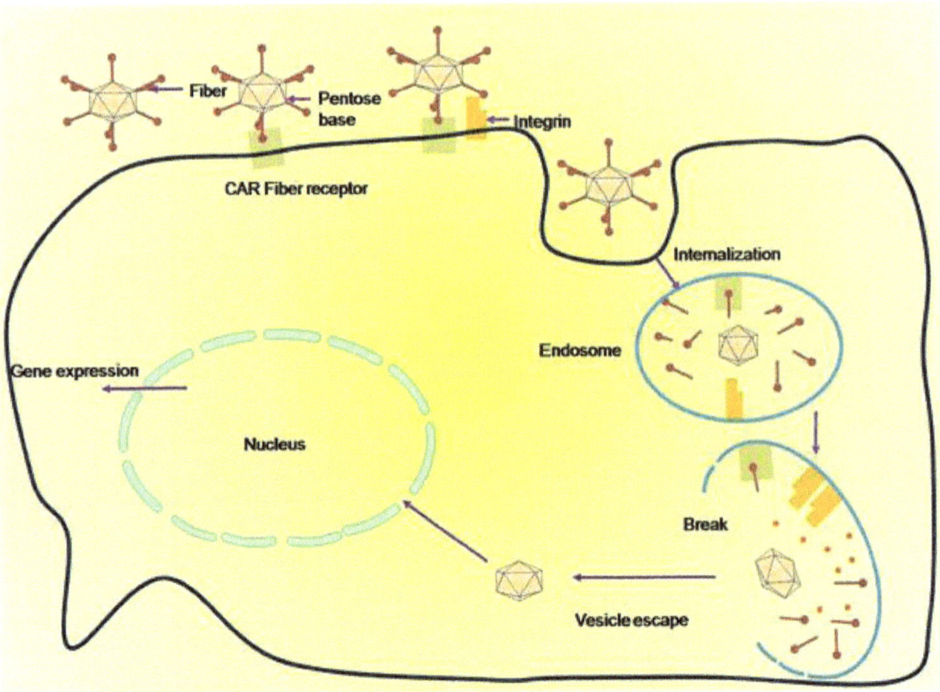

Adenovirus Associated Diseases

Respiratory Diseases

In young children and infants it causes an acute febrile upper respiratory tract infection. It may progress to pneumonia and pharyngeal infection in untreated cases and in immuno-compromised individuals. In adults the symptoms include fever with pneumonia and pharyngitis.

Latency inherpes viruses

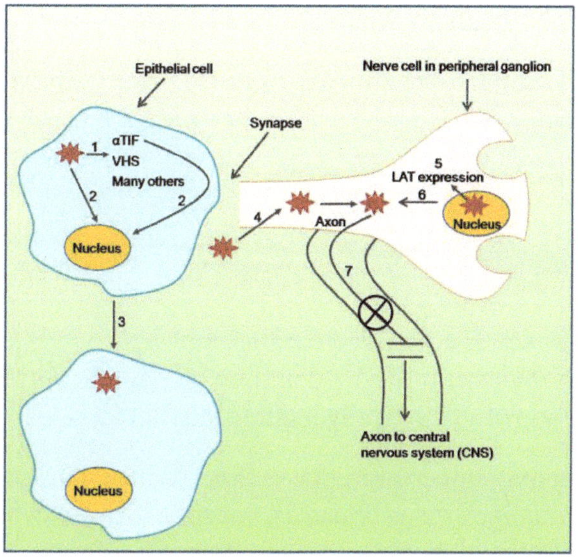

Other Diseases

In children the virus can cause acute gastroenteritis and hemorrhagic cystitis (inflammation of urinary bladder). Occasionally they may cause condition like meningoencephalitis in immuno-compromised patients. Sometimes they infect liver and eye leading to hepatitis and keratoconjunctivitis, respectively.

Adenovirus Pathogenesis

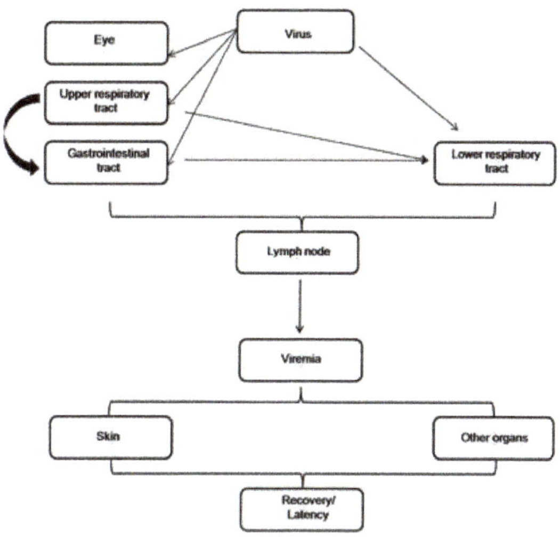

Prevention and Control

- Currently no vaccine is available to protect against adenoviruses.
- Good hygienic practices can prevent the infection.
- Hand washing is still the best way to avoid adenovirus infection.
- Wear protective clothing.
- Heat and bleach will kill adenoviruses.

Adenoviruses are unusually stable to chemicals, physical agents, and adverse pH, causing them to survive longer in environment.

Poxviridae

Poxviridae is a family of viruses. Human, vertebrates, and arthropods serve as natural hosts. There are currently 69 species in this family, divided among 28 genera, which are divided into two subfamilies. Diseases associated with this family include smallpox.

Four genera of poxviruses may infect humans: orthopoxvirus, parapoxvirus, yatapoxvirus, molluscipoxvirus. Orthopox: smallpox virus (variola), vaccinia virus, cowpox virus, monkeypox virus; Parapox: orf virus, pseudocowpox, bovine papular stomatitis virus; Yatapox: tanapox virus, yaba monkey tumor virus; Molluscipox: molluscum contagiosum virus (MCV). The most common are vaccinia (seen on Indian subcontinent) and molluscum contagiosum, but monkeypox infections are rising (seen in west and central African rainforest countries).

Structure

A) Electron micrograph of poxvirus particles in synovium of a big brown bat, northwestern United States. B) Negative staining of poxvirus particles in cell culture supernatant. Scale bar = 100 nm.

Poxviridae viral particles (virions) are generally enveloped (external enveloped virion- EEV), though the intracellular mature virion (IMV) form of the virus, which contains different envelope, is also infectious. They vary in their shape depending upon the species but are generally shaped like a brick or as an oval form similar to a rounded brick because they are wrapped by the endoplasmic reticulum. The virion is exceptionally large, its size is around 200 nm in diameter and 300 nm in length and carries its genome in a single, linear, double-stranded segment of DNA. By comparison, Rhinovirus is 1/10 as large as a typical *Poxviridae* virion.

Genus	Structure	Symmetry	Capsid	Genomic Arrangement	Genomic Segmentation
Betaentomopoxvirus	Ovoid		Enveloped	Linear	Monopartite
Yatapoxvirus	Brick-shaped		Enveloped	Linear	Monopartite

Cervidpoxvirus	Brick-shaped		Enveloped	Linear	Monopartite
Gammaentomopoxvirus	Ovoid		Enveloped	Linear	Monopartite
Leporipoxvirus	Brick-shaped		Enveloped	Linear	Monopartite
Suipoxvirus	Brick-shaped		Enveloped	Linear	Monopartite
Molluscipoxvirus	Brick-shaped		Enveloped	Linear	Monopartite
Crocodylidpoxvirus	Ovoid		Enveloped	Linear	Monopartite
Alphaentomopoxvirus	Ovoid		Enveloped	Linear	Monopartite
Capripoxvirus	Brick-shaped		Enveloped	Linear	Monopartite
Orthopoxvirus	Brick-shaped		Enveloped	Linear	Monopartite
Avipoxvirus	Brick-shaped		Enveloped	Linear	Monopartite
Parapoxvirus	Ovoid		Enveloped	Linear	Monopartite

Replication

Replication of the poxvirus involves several stages. The first thing the virus does is to bind to a receptor on the host cell surface; the receptors for the poxvirus are thought to be glycosaminoglycans (GAGs). After binding to the receptor, the virus enters the cell where it uncoats. Uncoating of the virus is a two step process. Firstly the outer membrane is removed as the particle enters the cell; secondly the virus particle (without the outer membrane) fuses with the cellular membrane to release the core into the cytoplasm. The pox viral genes are expressed in two phases. The early genes encode the non-structural protein, including proteins necessary for replication of the viral genome, and are expressed before the genome is replicated. The late genes are expressed after the genome has been replicated and encode the structural proteins to make the virus particle. The assembly of the virus particle occurs in five stages of maturation that lead to the final exocytosis of the new enveloped virion. After the genome has been replicated, the immature virion (IV) assembles the A5 protein to create the intracellular mature virion (IMV). The protein aligns and the brick-shaped envelope of the intracellular enveloped virion (IEV). These IEV particles are then fused to the cell plasma to form the cell-associated enveloped virion (CEV). Finally, the CEV encounters the microtubules and the virion prepares to exit the cell as an extracellular enveloped virion (EEV). The assembly of the virus particle occurs in the cytoplasm of the cell and is a complex process that is currently being researched to understand each stage in more depth. Considering the fact that this virus is large and complex, replication is relatively quick taking approximately 12 hours until the host cell dies by the release of viruses.

The replication of poxvirus is unusual for a virus with double-stranded DNA genome (dsDNA) because it occurs in the cytoplasm, although this is typical of other large DNA viruses. Poxvirus encodes its own machinery for genome transcription, a DNA dependent RNA polymerase, which makes replication in the cytoplasm possible. Most dsDNA viruses require the host cell's DNA-dependent RNA polymerase to perform transcription. These host DNA are found in the nucleus, and therefore most dsDNA viruses carry out a part of their infection cycle within the host cell's nucleus.

Genus	Host Details	Tissue Tropism	Entry Details	Release Details	Replication Site	Assembly Site	Transmission
Betaentomopoxvirus	Lepidoptera insects; orthoptera insects	Hemocytes; fat bodies	Glycosaminoglycans	Lysis; budding; occlusion	Cytoplasm	Cytoplasm	Unknown
Yatapoxvirus	Monkeys; baboons	None	Glycosaminoglycans	Lysis; budding	Cytoplasm	Cytoplasm	Contact; insects
Cervidpoxvirus	Deer	None	Glycosaminoglycans	Lysis; budding	Cytoplasm	Cytoplasm	Unknown
Gammaentomopoxvirus	Insects	Hemocytes; fat bodies	Glycosaminoglycans	Lysis; budding; occlusion	Cytoplasm	Cytoplasm	Unknown
Leporipoxvirus	Lagomorph; squirrels	None	Glycosaminoglycans	Lysis; budding	Cytoplasm	Cytoplasm	Arthropods; contact
Suipoxvirus	Swine	None	Glycosaminoglycans	Lysis; budding	Cytoplasm	Cytoplasm	Unknown
Molluscipoxvirus	Humans	None	Glycosaminoglycans	Lysis; budding	Cytoplasm	Cytoplasm	Sex; contact
Crocodylidpoxvirus	Crocodiles	None	Glycosaminoglycans	Lysis; budding	Cytoplasm	Cytoplasm	Unknown
Alphaentomopoxvirus	Monkeys; baboons	None	Glycosaminoglycans	Lysis; budding; occlusion	Cytoplasm	Cytoplasm	Contact; insects
Capripoxvirus	Sheep; goat; cattle	None	Glycosaminoglycans	Lysis; budding	Cytoplasm	Cytoplasm	Arthropods; contact
Orthopoxvirus	Humans; mammals	None	Glycosaminoglycans	Lysis; budding	Cytoplasm	Cytoplasm	Variola virus; Respiratory; contact; zoonosis
Avipoxvirus	Birds	None	Glycosaminoglycans	Lysis; budding	Cytoplasm	Cytoplasm	Arthropods; aerosol
Parapoxvirus	Humans; mammals	None	Glycosaminoglycans	Lysis; budding	Cytoplasm	Cytoplasm	Zoonosis; contact

Evolution

The ancestor of the poxviruses is not known but structural studies suggest it may have been an adenovirus or a species related to both the poxviruses and the adenoviruses.

Based on the genome organisation and DNA replication mechanism it seems that phylogenetic relationships may exist between the rudiviruses (*Rudiviridae*) and the large eukaryal DNA viruses: the African swine fever virus (*Asfarviridae*), Chlorella viruses (*Phycodnaviridae*) and poxviruses (*Poxviridae*).

The mutation rate in these genomes has been estimated to be 0.9-1.2 x 10^{-6} substitutions per site per year. A second estimate puts this rate at 0.5-7 × 10^{-6} nucleotide substitutions per site per year. A third estimate places the rate at 4-6 × 10^{-6}.

The last common ancestor of the extant poxviruses that infect vertebrates existed 0.5 million years ago. The genus Avipoxvirus diverged from the ancestor 249 ± 69 thousand years ago. The ancestor of the genus Orthopoxvirus was next to diverge from the other clades at 0.3 million years ago. A second estimate of this divergence time places this event at 166,000 ± 43,000 years ago. The division of the Orthopox into the extant genera occurred ~14,000 years ago. The genus Leporipoxvirus diverged ~137,000 ± 35,000 years ago. This was followed by the ancestor of the genus Yatapoxvirus. The last common ancestor of the Capripoxvirus and Suipoxvirus diverged 111,000 ± 29,000 years ago.

An isolate from a fish - Salmon Gill Poxvirus - appears to be the earliest branch in the Chordopoxvirinae.

Smallpox

The date of the appearance of smallpox is not settled. It most likely evolved from a rodent virus between 68,000 and 16,000 years ago. The wide range of dates is due to the different records used to calibrate the molecular clock. One clade was the variola major strains (the more clinically severe form of smallpox) which spread from Asia between 400 and 1,600 years ago. A second clade included both alastrim minor (a phenotypically mild smallpox) described from the American continents and isolates from West Africa which diverged from an ancestral strain between 1,400 and 6,300 years before present. This clade further diverged into two subclades at least 800 years ago.

A second estimate has placed the separation of variola from Taterapox at 3000–4000 years ago. This is consistent with archaeological and historical evidence regarding the appearance of smallpox as a human disease which suggests a relatively recent origin. However, if the mutation rate is assumed to be similar to that of the herpesviruses the divergence date between variola from Taterapox has been estimated to be 50,000 years ago. While this is consistent with the other published estimates it suggests that the archaeological and historical evidence is very incomplete. Better estimates of mutation rates in these viruses are needed.

Taxonomy

The name of the family, *Poxviridae*, is a legacy of the original grouping of viruses associated with diseases that produced poxes in the skin. Modern viral classification is based on phenotypic characteristics; morphology, nucleic acid type, mode of replication, host organisms, and the type of disease they cause. The smallpox virus remains as the most notable member of the family.

The species in the subfamily Chordopoxvirinae infect vertebrates and those in the subfamily Entomopoxvirinae infect insects. There are 10 recognised genera in the Chordopoxvirinae and 3 in the Entomopoxvirinae. Both subfamiles also contain a number of unclassified species for which new genera may be created in the future. Cotia virus is an unusual virus that may belong to a new genus.

Notes

The GC-content of these genomes differs considerably. Avipoxvirus, Capripoxvirus, Cervidpoxvirus, Orthopoxvirus, Suipoxvirus, Yatapoxvirus and one Entomopox genus (Betaentomopoxvirus) along with several other unclassified Entomopoxviruses have a low G+C content while others - Molluscipoxvirus, Orthopoxvirus, Parapoxvirus and some unclassified Chordopoxvirus - have a relatively high G+C content. The reasons for these differences are not known.

Phylogenetic analysis of 26 Chordopoxviruses genomes has shown that the central region of the genome is conserved and contains ~90 genes. The termini in contrast are not conserved between species. Of this group Avipoxvirus is the most divergent. The next most divergent is Molluscipoxvirus. Capripoxvirus, Leporipoxvirus, Suipoxvirus and Yatapoxvirus genera cluster together: Capripoxvirus and Suipoxvirus share a common ancestor and are distinct from the genus Orthopoxvirus. Within the Othopoxvirus genus Cowpox virus strain Brighton Red, Ectromelia virus and Monkeypox virus do not group closely with any other member. Variola virus and Camelpox virus form a subgroup. Vaccinia virus is most closely related to CPV-GRI-90.

Vaccinia Virus

The prototypial poxvirus is vaccinia virus, known for its role as the active agent in the eradication of smallpox. The vaccinia virus is an effective tool for foreign protein expression, as it elicits a strong host immune-response. The vaccinia virus enters cells primarily by cell fusion, although currently the receptor responsible is unknown.

Vaccinia contains three classes of genes: early, intermediate and late. These genes are transcribed by viral RNA polymerase and associated transcription factors. Vaccinia replicates its genome in the cytoplasm of infected cells, and after late-stage gene expression undergoes virion morphogenesis, which produces IMV contained within an envelope membrane. The exact origin of the envelope membrane is still unknown. The IMV is then transported to the Golgi apparatus where it is wrapped with an additional two membranes, becoming the Intracellular Enveloped Virus (IEV). The IEV is transported along cytoskeletal microtubules to reach the cell periphery, where it fuses with the plasma membrane to become the Cell-associated Enveloped Virus (CEV). This triggers actin tails on cell surfaces or is released as EEV.

History

Diseases caused by pox viruses, especially smallpox, have been known about for centuries. One of the earliest suspected cases is that of Egyptian pharaoh Ramses V who is thought to have died from smallpox circa 1150 years BCE. Smallpox was thought to have been transferred to Europe around the early 8th century and then to the Americas in the early 16th century. It is widely accepted that the main defeat of the Aztecs was due to a smallpox epidemic and within two years over 3.2 million Aztecs died. This death toll can be attributed to the American population's complete lack of exposure to the virus over millennia. A century after Edward Jenner showed that the less potent cow pox could be used to effectively vaccinate against the more deadly smallpox, a worldwide effort to vaccinate everyone against smallpox began with the ultimate goal to rid the world of the plague-like epidemic. The last case of endemic smallpox occurred in Somalia in 1977. Extensive searches over two years detected no further cases, and in 1979 the World Health Organization (WHO) declared the disease officially eradicated. In 1986, all virus samples were destroyed or transferred to two approved WHO reference labs: at the headquarters of the federal Centers for Disease Control and Prevention (the C.D.C.) in Atlanta, Georgia (the United States) and at the Institute of Virus Preparations in Moscow. Post September 11, 2001 the American and UK governments have had increased concern over the use of smallpox, or a smallpox like disease, in bio-terrorism.

Miscellaneous Viruses

Infectious diseases have played a s ignificant role throughout the history of mankind . Investigation of diseases dates bac k to ancient times and the query to understand it through science has lead to the discovery of viruses and bacteria as the causative agents of various types of infection and illness. Pathogenicity of viruses and susceptibility of host to infectious agents have constantly appeared throug h the emergence of new diseases and reappearance of pre-existing diseases. Emerging infectious viral diseases are those that have recently appeared in a population as a result of a new virus or the recognition of a previously undetected virus and are often zoonotic . Emergence of an infectious viral disease may occur due to the extension of the geographic or host range of the virus. Recently, bats have been implicated as an important reservoir and source of many emerging viruses. As new technology for detection of viruses becomes increasingly available, more viruse s are likely to be detected. E nhanced molecular biology techniques will allow faster and more com plete characterization of new and miscellaneous viruses.

Bat Paramyxoviruses

Bats have been shown to be the reservoir hosts of a variety of viruses responsible for severe disease outbreaks in humans and animals, including filoviruses, coronaviruses and paramyxoviruses . Recently Hendra and Nipah viruses were also isolated from bats.

Hendra and Nipah viruses are zoonotic viruses of the genus *Henipavirus* under the family *Paramyxoviridae*. The natural reservoirs for both the viruses are fruit bats or flying foxes of the genus *Pteropus* . Hendra virus was first isolated from an acute febrile illness in horses and subsequently in humans with a sign of fatal encephalitis. The first known human infections with Nipah virus were detected during an outbreak of severe febrile encephalitis in peninsular Malaysia and Singapore .

Menangle virus was isolated during an outbreak of reproductive disease in pigs in New South Wales, Australia in 1997. Symptoms of the disease included malaise, chills, fever, sweating, headache, weight loss and decrease in farrowing rate (birth giving process in pigs).

Tioman virus was isolated from the urine of fruit bats (*Pteropus hypomelanus*) . Tioman virus is lethal in suckling mice 8-12 days post intracerebral inoculation . There role in human and animal infection is still under debate.

Mapuera virus was isolated from the salivary glands of an asymptomatic fruit bat (Sturnira lilium) in 1979 in Brazil.

Canine Distemper Virus

Canine distemper virus is an important pathogen which naturally infects a broad range of terrestrial and marine carnivores. Canine distemper virus is a member of genus *Morbillivirus* of the family

Paramyxoviridae . The disease is characterized by skin rash, fever, gastrointestinal and respiratory signs, and a profound immune-suppression as well as by frequent neurological complications.

Rift Valley Fever Virus

Rift Valley Fever Virus (RVFV) is a member of the genus *Phlebovirus* (family *Bunyaviridae*). The RVFV is transmitted by the bite of mosquitoes. The disease was first reported in sheep in Kenya in 1918. Infection of RVFV is characterized by febrile illness with hemorrhages and inflammation of brain.

Hantavirus

Hantavirus belongs to the family *Bunyaviridae* (negative-sense, single-stranded RNA viruses) . Hantavirus is transmitted by rodents (deer mice) via their urine and feces. Hantavirus is a cause of hemorrhagic fever with a renal (Kidney) syndrome. The early symptoms of the disease are similar to flu and include fever, chills, cough and muscle ache. The disease can progress to Hantavirus pulmonary syndrome.

Ebola Virus

Ebola virus belongs to family of RNA viruses called the *Filoviridae* . The virus leads to fatal hemorrhagic disease in humans and nonhuman primates. The transmission of the virus occurs by direct contact with the blood and/or secretions of an infected person. Sudden onset of illness is characterized by fever, sore throat, headache, joint and muscle pain, and weakness, followed by diarrhea, vomiting, and abdominal pain. In highly fatal cases internal and external bleeding may be seen in the patients.

Arenaviruses

The family *Arenaviridae* contains the viruses which are usually associated with rodent-transmitted disease in humans. The virus particles are spherical with a diameter of around 110-130 nm. The virus contains negative strand RNA as a genetic material. Infection of Arenaviruses leads to hemorrhagic disease in humans that are often fatal.

Different Arenavirus diseases:

Virus	Disease
Junin virus	Argentine hemorrhagic fever
Lassa virus	Lassa fever
Guanarito virus	Venezuelan hemorrhagic fever
Machupo virus	Bolivian hemorrhagic fever

Prions

Prions are the infectious agents made up of only proteins (No DNA or RNA) and were discovered by Stanley Prusiner. Prions are propagated by transmitting the misfolded form of the protein. Prion diseases or transmissible spongiform encephalopathies (TSEs) are a family of rare progressive neurodegenerative disorders of humans and animals. They are characterized by long incubation periods, spongiform changes in brain, and a failure to induce inflammatory response.

Table Different prion diseases in human and animals:

Human Prion disease	Animal Prion disease
Creutzfeldt- Jakob Disease (CJD)	Scrapie
Kuru	Mad Cow Disease (Bovine Spongiform Encephalopathy)
Gerstmann- Straussler- Scheinker Syndrome	Chronic Wasting Disease
	Transmissible mink encephalopathy

Viroids

Viroids are plant pathogen that contains circular single stranded RNA as a genetic material. They are discovered by Theodor Diener in 1971. Viroids contain small RNA of around 250 to 500 nt and do not encode any proteins. The Potato spindle tuber viroid was the first viroid to be identified.

References

- Estep, R. D.; Hansen, S. G.; Rogers, K. S.; Axthelm, M. K.; Wong, S. W. (2012). Journal of Virology. 87 (1): 512–523. PMC 3536378 . PMID 23097433. doi:10.1128/JVI.02194-12
- Sandri-Goldin RM, ed. (2006). Alpha Herpesviruses: Molecular and Cellular Biology. Caister Academic Press. ISBN 978-1-904455-09-7
- Staras SA, Dollard SC, Radford KW, Flanders WD, Pass RF, Cannon MJ (November 2006). "Seroprevalence of cytomegalovirus infection in the United States, 1988–1994". Clin. Infect. Dis. 43 (9): 1143–51. PMID 17029132. doi:10.1086/508173. Retrieved 2009-12-04
- Huff J, Barry P (2003). "B-Virus (Cercopithecine herpesvirus 1) Infection in Humans and Macaques: Potential for Zoonotic Disease". Emerg Infect Dis. 9 (2): 246–50. PMC 2901951 . PMID 12603998. doi:10.3201/eid0902.020272
- Origgi FC, Tecilla M, Pilo P, Aloisio F, Otten P, Aguilar-Bultet L, Sattler U, Roccabianca P, Romero CH, Bloom DC, Jacobson ER (2015PLoS ONE. 10 (8): e0134897. PMC 4526542 . PMID 26244892. doi:10.1371/journal.pone.0134897
- John Carter; Venetia Saunders. Virology, Principles and Applications. John Wiley & Sons. ISBN 978-0-470-02386-0
- Chayavichitsilp P, Buckwalter JV, Krakowski AC, Friedlander SF (April 2009). "Herpes simplex". Pediatr Rev. 30 (4): 119–29; quiz 130. PMID 19339385. doi:10.1542/pir.30-4-119
- Whitley RJ (1996). Baron S; et al., eds. Herpesviruses. in: Baron's Medical Microbiology (4th ed.). Univ of Texas Medical Branch. ISBN 0-9631172-1-1
- Davison, A.J. (2010). "Herpesvirus systematics". Veterinary Microbiology. Elsevier. 143 (1): 52–69. PMC 2995426 . PMID 20346601. doi:10.1016/j.vetmic.2010.02.014
- Spencer, Juliet; et al. (February 2002). "Potent Immunosuppressive Activities of Cytomegalovirus- Encod-

ed Interleukin-10". Journal of Virology. 76 (3): 1285–1292. PMC 135865 . PMID 11773404. doi:10.1128/JVI.76.3.1285-1292.2002

- Murray PR, Rosenthal KS, Pfaller MA (2005). Medical Microbiology (5th ed.). Elsevier Mosby. ISBN 978-0-323-03303-9
- Spencer, Juliet (2007). "The Cytomegalovirus Homolog of Interleukin-10 Requires Phosphatidylinositol 3-Kinase Activity for Inhibition of Cytokine Synthesis in Monocytes". Journal of Virology. 81 (4): 2083–2086. PMC 1797587 . PMID 17121792. doi:10.1128/JVI.01655-06
- Weigler BJ (February 1992). "Biology of B virus in macaque and human hosts: a review". Clinical Infectious Diseases. 14 (2): 555–67. PMID 1313312. doi:10.1093/clinids/14.2.555
- Fenner, Frank J.; Gibbs, E. Paul J.; Murphy, Frederick A.; Rott, Rudolph; Studdert, Michael J.; White, David O. (1993). Veterinary Virology (2nd ed.). Academic Press, Inc. ISBN 0-12-253056-X

Permissions

All chapters in this book are published with permission under the Creative Commons Attribution Share Alike License or equivalent. Every chapter published in this book has been scrutinized by our experts. Their significance has been extensively debated. The topics covered herein carry significant information for a comprehensive understanding. They may even be implemented as practical applications or may be referred to as a beginning point for further studies.

We would like to thank the editorial team for lending their expertise to make the book truly unique. They have played a crucial role in the development of this book. Without their invaluable contributions this book wouldn't have been possible. They have made vital efforts to compile up to date information on the varied aspects of this subject to make this book a valuable addition to the collection of many professionals and students.

This book was conceptualized with the vision of imparting up-to-date and integrated information in this field. To ensure the same, a matchless editorial board was set up. Every individual on the board went through rigorous rounds of assessment to prove their worth. After which they invested a large part of their time researching and compiling the most relevant data for our readers.

The editorial board has been involved in producing this book since its inception. They have spent rigorous hours researching and exploring the diverse topics which have resulted in the successful publishing of this book. They have passed on their knowledge of decades through this book. To expedite this challenging task, the publisher supported the team at every step. A small team of assistant editors was also appointed to further simplify the editing procedure and attain best results for the readers.

Apart from the editorial board, the designing team has also invested a significant amount of their time in understanding the subject and creating the most relevant covers. They scrutinized every image to scout for the most suitable representation of the subject and create an appropriate cover for the book.

The publishing team has been an ardent support to the editorial, designing and production team. Their endless efforts to recruit the best for this project, has resulted in the accomplishment of this book. They are a veteran in the field of academics and their pool of knowledge is as vast as their experience in printing. Their expertise and guidance has proved useful at every step. Their uncompromising quality standards have made this book an exceptional effort. Their encouragement from time to time has been an inspiration for everyone.

The publisher and the editorial board hope that this book will prove to be a valuable piece of knowledge for students, practitioners and scholars across the globe.

Index